DONNIE ALLISON
AS I RECALL...

DONNIE ALLISON
AS I RECALL...

DONNIE ALLISON
WITH JIMMY CREED

Foreword by
Larry McReynolds

SPORTS
PUBLISHING

10 9 8 7 6 5 4 3 2 1

Allison, Donnie, 1939-
  Donnie Allison : as I recall... / Donnie Allison with Jimmy Creed ; foreword by Larry McReynolds.
     pages cm.
  Hardcover ISBN 978-1-61321-351-3 (hardcover : alk. paper) 1. Allison, Donnie, 1939- 2. Stock car drivers--United States--Biography. 3. Stock car racing--United States. I. Title.
  GV1032.A33A3 2013
  796.72092--dc23
  [B]
                               2013007415

Cover design by Brian Peterson
Cover photo by ISC Images & Archives

Paperback ISBN: 978-1-68358-489-6
eBook ISBN: 978-1-61321-439-8

Printed in the United States of America

To all my fans who I recall cheering me on whether I was winning or not.

To all the people who I recall helping me throughout my racing career by their sponsorship, friendship, team work, or just plain hard work.

To my family, who I recall always standing by me, not only as a celebrity, but as just a regular person.

To all the aspiring young drivers I have helped and will help in their climb up the ladder of big-time racing so that one day I will recall being a small part of helping them reach their dreams.

CONTENTS

ACKNOWLEDGMENTS

When you've piled up as many miles as I have over the years—both on and off the track—you meet a lot of people, make a lot of friends, and get a lot of help along the way. It makes for a lot of thank yous to say when the time comes, and to each and every one of you, I'm grateful.

In particular, I want to thank my family, my parents the late "Pop" and "Kittie" Allison, Bobby, Eddie and all my brothers and sisters, and my wife, Pat, and her parents, Joyce and the late Joe Leserra, and all our children for standing by me through thick and thin.

Thanks to my dad and mom for being the best parents a man could ever ask for and for not saying no when I was bitten by the racing bug. Thanks to Bobby and Eddie for all their support and encouragement over the years and for everything they taught me about racing. Thanks to Pat, Kenny, Pam, Ronald, Donald, and all the grandkids and now great-grandkids for being so understanding, so supportive and so loving for so, so long.

Thanks to my favorite uncle, Jake Allison, for the fly-fishing rod when I was young and for helping to instill in me a love of the outdoors at a very early age.

Thanks to Johnny and Jimmy Ardis, Clyde Bolton, Bo Brady, Jack and Betty Jo Davis, Dice Draper, Kenny Economy, "Hoss" Ellington, and A.J. Foyt.

Thanks to Bo Freeman, Horace and Nell Gray, Bob and Robert Hamke, Ralph Harris, Frank and T.D. Howton, Chester Honeycutt and J.W. Hunt.

Thanks to Charles LeCroy, Dr. Bill and Sue Lightfoot, Chuck Looney, Johnny Manfredi, "Banjo" Matthews and Ralph Moody.

Thanks to NASCAR, Dr. Jerry Petty, "Runt" Pittman, Ernie Reeves, Wayne and Donnie Russell, Jim and Janet Seaman, Jim Stacy, Ralph Stark, and Karl Sundman.

Thanks to Jon Thorne, Bobby Tindle, H.C. Wilcox, Ray Winston, Glen and Leonard Wood, Bob Wright and "Smokey" Yunick.

At this point I would also like to give special thanks to two outstanding men of medicine who I can honestly say helped save me after I was diagnosed with Hepatitis C in 2000, Dr. Robert Reindollar and Dr. Yut Sukkasem.

Hepatitis C is a contagious disease caused by a virus that infects the liver, and the Hepatitis C virus (HCV) is the most common cause of chronic liver disease in the United States and is increasing worldwide. Discovered in 1989, the World Health Organization now estimates between 130 and 170 million people are chronically infected with the HepC virus, with more than 350,000 of those dying from HepC-related liver diseases each year.

So you can imagine how earth shaking it was for me and Pat when it was discovered, out of the blue as you'll read late in this book, that I had this deadly disease. But due to the caring, persistence and professionalism of these two physicians, I was able to win my roughest race ever. There are not enough words in the English language to express the appreciation you have when you truly owe someone your life, but I hope these few will give these men some small idea of the special place I will always hold for them in my heart.

Finally, to anyone I should have recognized but can't quite recall—I'm sure there are plenty—thanks for all you've done to make it such a great ride. I appreciate you, one and all, more than you'll ever know.

— Donnie Allison

I want to thank all the folks over the years who believed I could write a book even when I didn't—especially the greatest college English professor ever, Gloria Whitten. Thanks, Glo, for staying after me to quit talking about it and just do it until the message finally got through my thick skull. And thanks for your time, your effort, your proofreading and copy-editing skills and for the constant refresher course in the English language.

Thanks to my parents, Shelton and Shelby Creed, who heard me say "I'm working on the book today" so many times over so many months, I'm sure they got sick of it, but who hung in there with me anyway.

To the countless friends and colleagues who heard me say the same thing time and again and who still count me as a friend and colleague.

To Jack Davis, good ole' Jack P. as Donnie calls him, and his wife, Betty Jo, for opening their home and their hearts years ago and making me feel like one of their own.

To my Biloxi buddy, Angie Balius, for providing a non-race fan's point of view and being candid about what was good and what wasn't in that respect.

To Betty Carlan, the former librarian at the International Motorsports Hall of Fame in Talladega, Ala., for her help in researching this subject.

To Kristi King, former media relations director at the Talladega Superspeedway, for loaning me her racing history books for long stretches of time, and to her assistant at that time, Lisa Battles, for her help with research as well.

To Lynnette Bogard for taking a chance on someone who had never written a book before way back in 2004 and being patient while I learned.

Finally, thanks to Donnie and Pat Allison for opening up to me and allowing me to tell this story. It is a privilege and an honor I can't put into words.

— Jimmy Creed

FOREWORD

In the early spring of 1980, I was at a crossroads in my racing career. I had an offer on the table to leave my home in Birmingham, Ala., and join a new NASCAR team being formed in Greenville, S.C. It was a good offer, and I was pretty sure I wanted to do it, but I was still going back and forth on my decision.

So I picked up the phone, called Donnie Allison at his shop in Hueytown, Ala., and asked if I could stop by for a chat. He and a man named Bo Brady were there working on his car and invited me to come on by. When I got to the shop, Donnie stopped what he was doing even though he was behind on getting things ready and listened to what I had to say.

After I'd laid it out for him, I'll never forget what he did. He walked up, put his finger right in my face and said, "You need to do it." Then he gave me a piece of advice that I still pass along to others who ask me about getting into the racing business to this day.

Donnie told me to go get the checkered flag from the last race I'd won at the Birmingham International Raceway with a driver named Mike Alexander and put it up where I could see it because it was going to be a long, long time before I got another one. And he was right.

I went to work for Rogers Leasing Racing in 1980 and it was late 1988 before I finally got to go to victory lane after winning my first Winston Cup race.

As I walked into victory lane that day at Watkins Glen after Ricky Rudd had won the race, one of the first people I thought about was Donnie. I thought about just how dead on he had been with his advice to me. How I had truly had to work harder to lose races at that level than I ever did to win them on the short tracks. And how much I respected him for sharing it.

Donnie wasn't trying to discourage me or tell me not to do it that day. I'm pretty sure he knew how badly I wanted to do it and where I wanted to go. Donnie just told it like it was and gave me a very, very clear picture of how hard I was going to have to work to get there.

Anybody who knows Donnie isn't surprised that he shot straight with me that day. It's what you expect when you deal with him.

No truer scouting report has ever been written than the one that labeled Donnie Allison as a man who tells it like it is.

The thing that made me respect him so much over 30 years ago and still today is that, if you're talking to Donnie just because you're looking for somebody to massage your ears in reply, you're talking to the wrong guy.

If you ask Donnie for his opinion, be prepared to hear exactly what he thinks. It may not be what you thought you'd hear or what you want to hear, but it'll definitely be what he thinks you need to hear.

To be honest, that approach to life and to racing my have worked against him as a driver and been part of the reason he never got the acclaim that a lot of other drivers got.

I honestly believe Donnie was as good a driver as Richard Petty, Bobby Allison, David Pearson, Cale Yarborough, Darrell Waltrip, all of them. He just never got the breaks those guys got that put him in a position to be ultra-successful like they were.

Part of the reason for that, I believe, is because he was so candid. If ever a person was too honest in our sport, it was Donnie Allison. He was there to do a job driving a racecar and if you were looking for somebody to smile, shake hands and make nice, you were probably talking to the wrong guy.

The same goes for this book. If anyone expects him to reflect back on his career, to tell the stories and share the memories, and not do it in a completely straightforward fashion, you're talking about the wrong guy.

Anybody who knows Donnie knows the things he has related here are told exactly as he remembers them, uncut and uncensored, and if that steps on any toes, well that's just the way it is. Anybody who knows him wouldn't expect anything less.

Contrary to what many may think, I never worked for Donnie, Bobby or Davey during my short-track days in Birmingham. But I was fortunate to develop extremely close relationships with them all later, relationships that are now some of the most cherished of my entire life.

Two other things come immediately to mind when I think about my friendship with Donnie over the years.

The first happened during the fall Winston Cup race in Phoenix, Ariz., in 1991. I was crew chief for Davey with Robert Yates and we were so bad in the final practice on Saturday that we were off the charts.

Donnie had flown out with Davey that weekend and he pulled us aside, ran some things by us and with what seemed like only about five minutes to go in practice, we put some stuff under the car that Donnie had thrown at us. What he told us turned that racecar around and we went on to win the race the next day.

The other thing I remember vividly about Donnie was his driving that black No. 28 around Talladega that day in July 1993 when we paid tribute to Davey. That had to be one of the hardest things he ever did in his life because he loved Davey like a son.

But Donnie never backed away from the hard things or hard work and that's what made him one of the best ever in my book.

When he asked me to do write this introduction, I was absolutely blown away. When I think of all the people Donnie has come in contact with over the years through racing, all the friends and acquaintances he's made, I'm sure it's an infinite number. So for him to choose me to do this for him from among all those was the absolute highest of honors and a request I could not refuse.

Way back in 1980, Donnie Allison sent me in pursuit of my dream with the simple instruction to be prepared to work harder than I ever had in my life to achieve it. Now that I have, I can't thank him enough for it.

As I said, Donnie was one of the first people I thought of when I finally got to touch my first Winston Cup checkered flag in 1988. As we celebrated amid the champagne shower, one other thing I remember thinking was that almost nine years ago, Donnie had told me it was going to take that long and, man, was he right. As usual.

— *Larry McReynolds*

Devastated in Daytona

The rain splattered hard against the hotel room window. As I stood watching the storm rolling across Daytona Beach, I kept saying under my breath, "This is not good. This is not good."

That happened the night before a lot of my races, though that might come as a surprise to many. Throughout my racing career, I had the reputation—even within my own family—of being the Allison that didn't care. But I did care. A lot more than I ever showed, I guess.

I wanted to win races, and I especially wanted to win the next day's race, the 1979 Daytona 500.

It was every stock-car driver's dream to win the "Great American Race" on the high, fast banks of the Daytona International Speedway. I was about to have my best shot, if the weather didn't ruin it.

I had plenty of reasons to be excited.

I had qualified second, giving me a front-row starting spot alongside pole sitter Buddy Baker. I had a fast car. My car owner, "Hoss" Ellington, and my crew had done a great job getting the No. 1 Hawaiian Tropic Oldsmobile ready. Most of all, I was confident.

Give me a good car on a fast track like Daytona, and I usually had a chance to win.

The clouds and rain still hung over the track the next morning, but there was no way the race was going to be canceled, not with CBS Television cameras in place and set to make history.

The network was going to beam the race live from start to finish to homes across the country for the first time ever, and there was no way Bill France Jr., the iron-willed owner of NAS-CAR, was issuing any rainchecks.

We knew we were going to race. What we didn't know was that the weather was much worse in other parts of the country, so bad it had paralyzed almost the entire Eastern seaboard under a thick blanket of snow.

From Jacksonville, Fla., to Maine, folks found themselves stuck in their homes that Sunday with little else to do but flip channels on the TV.

Back then, there weren't a hundred channels to choose from like there are today. There were the three major networks, ABC, NBC and CBS, and if you didn't watch one of those, you didn't watch anything.

So the channel surfers that day came upon the Daytona 500. Since they'd never seen anything like it, and since they didn't have much choice, many watched. What they saw was the most memorable finish—and most significant race—in NASCAR history.

After what seemed like forever, we started engines. I really didn't get nervous as we waited for the crews to dry the track, but I was a little more apprehensive than usual. Mostly I was worried about someone else doing something stupid on a track that was still going to have wet spots on it when we started.

They decided to have us run caution laps so that all the cars circling the track could help dry it. Then they'd give us the green flag and we'd be off, and I'd get my first Daytona 500 victory.

I'd come close in 1974, but had blown a tire 12 laps from the finish with a 19-second lead over Richard Petty. On this day, I was going to get it. I could feel it.

Starting in the row behind me were Cale Yarborough and Darrell Waltrip. My brother, Bobby, was two rows behind that, and Richard Petty was even further back. By day's end, three of them would be directly involved in the bitterest disappointment of my life.

When it came time to start the race, a funny thing happened. The weather was bad all night and all morning, but as we pulled off pit road, the sun broke through the clouds. By the time we'd run 16 caution laps, it had turned into a nice, warm day.

I'd always said "Big" Bill France—Bill Jr.'s father—was a personal friend of the Man upstairs, and this was the perfect example. Here was the first live national telecast of a race. Half the country was snowed in, the other half wet, and he got sunshine for race day. He definitely had a connection.

The trouble started on lap 31. I was leading, Bobby was second and Cale was third as we came off turn two. Bobby pulled inside of me and got his right-front fender just past my left-rear bumper. Then Cale touched him, just a tap, on his rear bumper.

Bobby turned up into me, and I lost control. I spun in front of him, and he hit me just behind the driver's side door so hard it lifted three of my wheels off the track and almost sent me tumbling.

We went spinning off the track into infield grass down Daytona's long backstretch, and to avoid the wreck, Cale had to come right along with us. The infield was a swamp from all the rain, and we all got stuck in the mud.

As we sat spinning our wheels, the other cars circled the track, putting us laps down and taking away any chance we had of winning the race. Or so it seemed.

When we all finally got back on track, NASCAR had me two laps down and Cale three. Bobby's chances were gone, but he would figure in the finish. Man, would he ever.

The car coughed and sputtered some from water in the engine, but it actually handled better after the spin than before it. My car was in good shape and so was Cale's. That told me he'd be there at the finish.

We both went to work, and what followed was, simply, one of the greatest comebacks ever in a NASCAR race.

Today at Daytona, if a driver falls two laps down, it almost certainly dooms him since most of the 43 cars will finish within seconds of each other on the lead lap. In 1979, I made up five miles and Cale made up seven-and-a-half. My Oldsmobile was so strong it only took me 77 laps to do it, too.

I swept back into the lead on lap 108 and kept a tight grip on it most of the rest of the way. Dale Earnhardt led a few laps,

Waltrip a few, A.J. Foyt a few. Even Cale led one late. But this was my race to win.

As I said, I worried about somebody doing something stupid that would affect my chances of winning. One of the drivers that concerned me was Cale.

It wasn't that he did stupid things. He just ran so hard to lead races and win races that drastic things could happen if you got in his way. The final lap was proof of that.

I took the white flag, two-and-a-half miles from the checkers, with Cale close on my tail.

We went into turn one, I looked in the mirror to see where he was and saw him going low. I made up my mind he was not going to get under me coming down the backstretch. He could have all the room he wanted on the outside, but he was not going to get under me on the back straightaway.

As we came around to turn two, I was in the third lane, and I moved about a half a car width down from the normal racing groove. Suddenly, he hit me in my back bumper, and when he did, I turned sideways a little bit.

I was almost off the corner anyway and I was a little bit sideways, so I lifted. When I lifted, he hit me in the door. That put him down on the apron, his two left wheels in the dirt.

When he came back up on the racetrack, he was turning right, and I was trying to get my car straightened out, and we hit again—this time pretty hard. That blow actually knocked him back down into the grass and me back up onto the track a little further.

The next time he came out, he was cutting the wheel to the right and standing on the gas as hard as he could. When we hit the third time, we really hit, and I said, "To hell with it. If we crash up here, we'll both be wide open."

He never backed off. He never tried to slow his car down. Well, we did crash, and Petty slipped by us to win his fourth Daytona 500.

We ended up in turn three, he got of his car, I got out of my car, and we had a conversation there, which was very unpleasant. We called each other a few unpleasant, unprintable names, and that was about it.

Our conversation was pretty well over when Bobby stopped on the apron and asked if I was all right and if I needed a ride back to the garage. All this time, Cale was walking over to Bobby's car. I didn't hear exactly what he said, but it was something

This confrontation between Cale Yarborough and me after the disastrous end to the 1979 Daytona 500 proved to be one of the milestone moments in NASCAR history. **(Courtesy of Donnie and Pat Allison)**

to the effect that it was Bobby's fault because he'd been block-ing all day long. Then Cale took a swing at Bobby through the window with his helmet and cut Bobby across the bridge off the nose, drawing blood.

I ran over and grabbed Cale and told him if he wanted to fight, I was the one he needed to fight with. Then here came Bobby. I could not believe that Bobby Allison got there that fast. But I had seen that look in his eyes before, and I knew what was going to happen. After that, as Bobby likes to say, Cale went to beating on Bobby's fist with his nose.

They had schooled us all week that the race was going to be on live TV. Live TV. Live TV. That's all we heard. Well, I didn't think about it one time during the race, but standing in turn three with Bobby and Cale going at it, it suddenly came to mind.

There's a famous photo of me holding Bobby's elbow. Well, I was reminding him that the race was on … live TV.

The TV viewers—and everybody else—got their money's worth that day. Bill France Jr., "Big" Bill's son, pulled me aside in the garage afterwards and told me NASCAR didn't need stuff like that to sell tickets. Oh really?

Well, I wish there was some way we could calculate how many tickets NASCAR has sold since the 1979 Daytona 500. Trust me, it would be astronomical how may tickets they've sold and how much money they've made because of that fight.

People often ask me if I'm bitter towards Cale Yarborough. I'm not bitter. I'm still just very, very disappointed in that day.

Disappointed because I know if I'd been second to Cale, I would not have done what he did. But I can't judge other people by me.

Cale and I are casual friends. We're not going to go out to dinner together, but we didn't before the race. I'm good friends with Betty Jo, Cale's wife. I speak to them everywhere we go.

If I had an opportunity to do something with other drivers, I would ask Cale to go, because maybe we'd talk about the wreck. Cale Yarborough and Donnie Allison have never talked to one another about the wreck other than that day in the infield after it happened, and all I said then was, "You crazy SOB."

I did say one time in front of him on national TV that he wrecked me. I said it then. I'll say it today and, if I'll live another 25 years, I'll say it in 25 years because that's what happened.

Cale Yarborough took away my best shot at winning the Daytona 500. Fair and square, he would never have beaten me.

A Jockey, a Pitcher, or a Diver

I didn't grow up wanting to be a racecar driver. I wanted to be a jockey. Or a star baseball pitcher. Or an Olympic diver.

As a kid in Miami, I was good at sports. We used to go to the Pan American Building on Delaware Parkway every day to play football and baseball. I was a little, tiny shrimp, but I was always the first one chosen, or they made me choose because I was usually on the winning team.

I wasn't much good at football because I was so small, but I could fire a baseball.

My older brother, Eddie, had been a bat boy for the Miami Sun Socks, a Brooklyn Dodgers farm team, and he had gotten us a pitching rubber and a home plate, so we built a mound. We built it 60 feet, six inches just like the pros, and that's how I practiced. I threw a ball 60 feet, six inches every day, even as a kid.

Well, when I went to play Little League, I only had to throw it 40 feet, so I could really bring it. I threw so hard they wouldn't let me throw batting practice. The coach on the Optimist Club team I played for told me one time he couldn't let me throw batting practice because nobody could hit it and that wasn't good for the team.

I was 14 years old in 1953, left-handed and could throw a curve, a drop, a fastball, and a slider. Too bad my coach at St. Leo's Catholic Boys School couldn't see it.

I played for the St. Leo's team all year, but for some reason the priest who was the coach would never let me pitch. So it was very surprising when, in the last inning of the last game of the season, he told me to go warm up.

We were playing our rival Dade City, and they had a guy on first and nobody out when the priest sent me to the bullpen. I was so excited I ran over and threw three or four pitches and told him I was ready.

The next guy got on base, so they had runners on first and second, nobody out, and the coach finally put me in the game to hold the lead.

I wasn't scared because I had really been looking forward to getting in there. Besides, I was young, bulletproof and not intimidated by anything.

I struck the first batter out. I struck the second batter out. Then came the third batter, who was their best hitter.

My first pitch to him the ump called a ball. He called the second one a ball, too, even though I felt sure it was a strike. On the third one, I threw my slider, which went towards the batter then away from him.

Well, he timed it, and all I heard was this loud crack when he hit it. I looked straight up over my head as the ball was leaving. I didn't even turn around. I felt sure it was gone.

But the centerfielder never took a step. He just put up his glove and caught it for the third out, and everybody started jumping up and down and hollering. When all the hoopla was finally over, the priest said to me, "I didn't know you could pitch like that."

Now I was only in the ninth grade, but I looked him straight in the eye and said, "Father, that's because you never gave me a chance."

That was the last game of the year, and I didn't go back to that school the next year. I went to a Catholic boys' school, Archbishop Curley High School, for part of my 10th-grade year before I dropped out. But I never tried out for the baseball team again.

I just didn't feel like they would let me make the team, or if I did, they wouldn't give me a chance to prove myself. That's all I ever wanted out of anything in my life. Just give me a chance to prove myself, and I've usually succeeded.

That was what was so appealing about swimming and diving and riding horses. It was something I could do on my own. I wasn't at the whim of a coach to put me in the game or a teammate to make the catch for the out. If I worked hard, I could make my own success, and I liked that.

When I started swimming and diving, I did it all by myself. I went to the Curtis Park public swimming pool by myself. I worked out by myself.

There was a lifeguard at the pool named Wilbur Shaw who took an interest in me and helped me some, but mostly I worked alone to make myself a champion.

When I was 13 and 14, I won all the Boy Scout meets. It was funny, but the troop leaders used to recruit me to swim for their troops like they do a kid to play college football or basketball today.

In 1953, I participated in the Boy Scouts championship in Miami, which was a pretty big meet. We had a lot of troops from all around town. I entered five swimming and diving events, and I won all five.

By 1954, I had advanced to the AAU level and decided to concentrate on diving off the 1-meter and 3-meter boards. I loved flipping, twisting, anything with a reverse layout—the higher the degree of difficulty the better.

I went to the Florida AAU championships in Largo, Fla. that year, got to the pool, took one look at the diving board and felt like somebody had punched me in my stomach.

At the municipal pool I worked out in back in Miami, we had gotten some new aluminum diving boards, and I loved them. I'd bend the fulcrum back a little bit and bounce on them like crazy, which was good for an 80-pound kid.

In Largo, they had a wooden slat diving board, which was like jumping on concrete.

The coach told me to go work out the first day. I walked over, looked at the board and didn't even get on it. I went back and told him I couldn't dive on that board. He said I had to because they didn't have another one.

I went up and tried a couple of dives and was really discouraged because I couldn't bounce that board. To be perfectly honest, I was very disturbed. I told myself I was going to be terrible the next day. Boy, was I wrong.

The next day, I hit every dive just like you would draw it up on paper. I don't know how or why it happened, but it did.

I annihilated a field of 28 divers in my age group. I got 8s and 9s on my dives when everybody else got 4s and 5s. I won by 27 points. Do you know how hard that was to do?

One of the coaches at the AAU meet was from Coral Gables High School, and he told me he thought if I would go to a

I had a determined look in my eye even as a young boy. (**Courtesy of Donnie and Pat Allison**)

public school—I assumed he meant his school—I could get a college scholarship. But I wanted to stay in Catholic school, so I declined.

While I was doing all this swimming and diving at the municipal pool, I met a man—I don't even remember his name now—who trained horses at Tropical Racetrack. He came to the pool because he liked to dive and wanted to talk diving, but I was more interested in talking about me riding racehorses than about his diving.

At that time, I was riding horses with a friend of mine, and I was taking care of a lady's horse for her, too. When I told him of my interest in horses, that my secret passion was to become a jockey and that I only weighed about 90 pounds, he told me to come look him up. So I went.

He told me I had to start somewhere, so he had me start by giving the horses some hay.

The next time, I went early in the morning, and he let me ride a lead pony, one they used to lead the racehorses out to the track. He could see I could ride, so he wasn't concerned about me.

Next, he let me get on a racehorse, not to ride it on the racetrack, but to walk it around the paddock area. I got on that horse, sitting there in the paddock area, and felt like I was on top of the world.

Finally, I was going to ride an actual racehorse on an actual racetrack. But I never got to do it. A 1948 Chrysler saw to that.

The Big Crash

I had saved my money, maybe 40 or 50 bucks, and bought this little motorcycle called a Mustang. It was a scooter really, but it had gears and three forward speeds. It was a good motorcycle, and that's how I got around town.

On March 23, 1956, a friend named Mike Hopkins and I were on the scooter going from Tropical Racetrack out to this lady's house where I took care of a horse. We were following a 1951 or '52 Ford pickup down Milam Dairy Road, just west of what is now the Miami International Airport.

All of a sudden, the driver jammed on his brakes, and I was left with a decision.

I couldn't stop; that was for sure. I could run into the back of the truck, turn right into a ditch or go left into the other lane. I cut to the left and clipped the taillight, which in those days was mounted with the license plate in a bracket on the left side of any Ford truck.

There was one car coming the other way, a 1948 Chrysler four-door sedan. It went by just as I clipped the taillight, which sent me flying into the Chrysler. I hit on the driver's side, where the back fender met the door.

After the impact, I was left lying in the middle of the road; the motorcycle was on the shoulder, and Mike was lying there, too. When I looked down at my feet, I knew I had a problem because I had a pair of cowboy boots on, and one was pointing up and the other down.

All I could think of was to get out of the middle of the road immediately. So I rolled over to the shoulder and ended up lying with my head downhill a little bit. I was still lying that way when some friends of mine got there to help me.

They rolled me over and found me bleeding so badly the blood had saturated my clothes and jacket and formed a puddle in the gravel beside my head. I had suffered a double compound fracture of my left leg, and whatever caught my knee opened a cut that took 188 stitches to close inside and out. It wasn't pretty then, and it doesn't look too good now.

When I got to the emergency room, I had lost so much blood they immediately gave me a blood transfusion. Then they started talking about amputating my leg, which at the time would have been the easiest thing to do.

They might have, too, but my father, Edmond "Pop" Allison, told them they were not going to take my leg.

They had me all hooked up to put me to sleep. They had given me a shot and told me to start counting backwards from 100. That's how close they were to starting the procedure to take my leg when the main orthopedic surgeon, Dr. McDonald, told them to wait because he wanted to check my leg again.

Dr. McDonald felt of my leg and said they couldn't operate because there was too much fever. When one of the bones came through the skin, it stuck in the asphalt, and I think that caused infection to set in. So in a way, that infection saved my leg because I was so feverish they couldn't operate.

I really don't know how they saved my leg. Nowadays it probably wouldn't be anything to reconstruct my leg, even if it had been broken that badly. But this was 1956, and they didn't have the technology to do anything like that. Honestly, I think they just experimented on it, and whatever they tried happened to work.

I never did have an operation on it. Instead, they put me in a room in traction, and I stayed in traction for 13-1/2 long weeks.

It was tough for a 16-year-old kid, especially one as active as me.

They finally let me go home, but I was in a body cast for eight more weeks after that. The cast ran from the middle of my chest down my left leg to my ankle. I didn't have anything on my right leg, but that was because I wasn't supposed to do anything. I was just supposed to lie in bed.

They didn't even give me any crutches when I left the hospital because they didn't want me up moving around. They didn't know Donnie Allison, though.

At that time, I was doing a lot of roller skating, and I told all my friends I would roller skate again before my birthday rolled around on September 7. On September 2, I roller skated again just like I said. I didn't skate too good, but I skated.

I'd only been out of the hospital maybe two weeks when my good friend Larry Owen left to go in the Air Force, and they gave him a going-away party. I was in a body cast. I didn't have any crutches. I wasn't supposed to be out of bed. But I was determined to go to that party.

I got down the street to Larry's house where they had set up a couch for me in the backyard. They got me situated, and I was enjoying the party when Larry's brother, James, came over and said he had some crutches.

I told him to go get'em. He gave them to me and I tried them. I hadn't done anything for a long time, but I got up and tried to walk around the backyard on those crutches. I barely made it back to the couch before I fell down. But I took those crutches home and kept trying to walk on them.

The funniest thing I did with my body cast on was convince my mother to let me drive a car.

My left leg was immobile, but the car was an automatic, so I only needed my right leg. I couldn't bend forward because of the cast around my waist, so I had to sit in the car seat like I was lying down. It wasn't very convenient, but it worked. I might not have been able to get out once I got where I was going, but at least I was up and moving enough to get there.

The accident changed the direction of my life significantly.

Before then, I had seriously considered being a jockey, which would have meant making my living using one horsepower instead of 600. But after I went in the hospital weighing 105 pounds and came out 13 weeks later weighing 125, the idea of being a jockey was gone.

I was actually supposed to ride a racehorse for the first time at Tropical Racetrack the day after my motorcycle wreck, but fate changed that plan. Later on, I did try to ride a racehorse, but my leg had been broken so badly I never could bend it all the way after that, and it was painful.

AS I RECALL...

The injury left my left leg three-fourths of an inch shorter than my right one. It left my spine crooked, and it gave me a limp I have to this day. Later, it sometimes made running 500-mile NASCAR races pretty painful, too.

Also, I missed school all those weeks I was laid up and never went back. But even though I went to work with my dad installing service station and garage equipment and he needed the help, I'm still not real proud of the fact I was the only one of 10 Allison children who did not graduate from high school.

The wrecked motorcycle sat in my father's garage out back of our house for a long time. When I finally got up and going again, I went out, took it apart and straightened it out. I fixed it up and sold it.

Of course, before I sold it, I had to be sure it was ready. The way I looked at it, I couldn't sell somebody damaged merchandise. That just wouldn't be right. So I got back on and tested it out with ... that's right ... my leg still in that cast.

Cast or no cast, nobody could keep this man down. Or off that motorcycle.

One Big, Happy Family

Edmond Jacob Allison and Katherine Allison—folks called her Kittie—were devout Catholics and had 13 children to prove it. Three of the children—two girls and one boy—died at birth or shortly after. But for much of my childhood, there were 10 children ripping and roaring around the Allison house.

For much of that time, we lived in a house in northwest Miami with just three bedrooms. That was on N.W. 19th Street between 31st and 32nd Avenues. When I was in the fifth grade, we moved to what we called "The Big House," and I couldn't believe there was that much room in the world.

It was a 17-room, two-story house on 15th Street between 30th and 31st Avenues that had five bedrooms and a bath upstairs and a master bedroom and another bath downstairs.

That still wasn't enough bathrooms, though, especially when the older girls got in there. If you didn't get in there before they did, you were out of luck. If you had a "necessity of life" at that point, you had to head downstairs because you're weren't getting in that upstairs bathroom.

Even with such minor inconveniences, my childhood was as good as it could have been.

We always had something to do. We never wanted for anything. We may not have had the luxuries that some people had, but we didn't even think about it.

We had a ball, a glove and a bat to play baseball. It might have been used, but it was as good as anybody else had. We all got fishing poles if we wanted them, too.

I remember my first fly fishing rod. Uncle Jake, my dad's only brother, was a big fly-rod fisherman, and he was my favorite, and I was his favorite. So, of course, I became a real fly-rod buff, too.

We hadn't even moved into "The Big House" yet when I got that first fly rod. I remember standing outside in the yard practicing by throwing my flies into a garbage-can lid turned upside down and filled with water. It's a wonderful memory.

Every one of us got a bicycle at Christmas. My dad got them at the police auction and re-did them, but to my recollection, every one of them were as good as brand new.

Another thing I can honestly say we got plenty of, even with there being 10 of us, was attention from our parents.

We had friends in our neighborhood whose fathers never came outside to play or throw a ball with their sons the whole time we knew them. My dad would.

As tired as he'd be from his job, he'd still come out and play catch. When I wanted to pitch, he caught for me. At least until the day I was 13 or 14 when he finally told me he wasn't going to catch me anymore because I burned his hand too badly when I threw.

Most of all, we had a lot of love in our house and that went a long way.

My oldest sister, Dorothy, was born in 1929, but she died at birth. After that, Claire was born in 1931, Pat in 1934, my oldest brother Eddie in 1936, Bobby in 1937, me on Sept. 7, 1939, Tommy in 1940, Stanley in 1942, Mary Catherine in 1944, Margaret in 1946, Jeannie in 1947, Mary Agnes in 1950 and Cindy in 1954.

Stanley died five days after he was born, and Mary Catherine was born on Jan. 5, 1944, and died on Dec. 8, 1944. Margaret also left us way too early, a victim of cystic fibrosis that took her when she was a junior in high school.

The thing I'll always remember about Margaret is that she was a fighter.

For one thing, the normal life span of a child with cystic fibrosis was four or five years, and she lived to be 16. Another amazing thing about her was that the year before she died, she had perfect attendance for her 10th-grade year at school. It was quite something for somebody that sick not to miss a day of school, and it made an impression on me.

As I got older, into my middle teens, and started going out dating and other things, I'd come home at night and hear Margaret coughing. It was one of the most helpless feelings in the world because I wanted to go help her, but there was nothing I could do.

The last year of her life, she slept under a tent. My dad elevated the head of her bed, and she slept under that tent so she wouldn't cough all night and get choked up. In a big houseful of healthy, active children, it was hard for us to see our sister suffer like that, and it probably had as great an effect on me as any of my childhood experiences.

Margaret's condition did help us understand a little bit better what took our other siblings, though. Until Margaret was diagnosed with cystic fibrosis, all of their deaths had been attributed to pneumonia or fluid on the lungs. Well, that's basically what cystic fibrosis is, so we figured all of them had a touch of CF.

With that many kids, my parents had their hands full already, but that didn't stop them from taking in others as well.

One or the other of my grandparents—my mother's father or my dad's mother—lived with us all the time. Plus, my mom often had a child from the Catholic orphanage, and we treated them just like family. One time we even had two kids from the orphanage, and we all truly acted like one big family.

In fact, we used to laugh and say that when we sat down for a meal at the Allison house, it looked like the Last Supper because so many people were there.

That's another thing I'll remember about my childhood. We always had plenty to eat.

One of my most vivid childhood memories is of getting up each Sunday morning and going as a family to 6 o'clock mass at the Gesu Church in downtown Miami. On the way home, we'd stop at August Brothers Bakery and my mother would buy hard rolls, soft rolls and onion rolls, and that's what we'd have for breakfast every Sunday morning—rolls, coffee and milk.

Then she'd fix Sunday dinner, and we'd all sit down and eat as a family. She did that every Sunday I can remember, and if you were there, you were expected to sit down and eat with the family.

It was the same way during the week. She fixed supper every night, and if you were home, you were expected to come to the table to eat. But buddy you didn't eat anything until you got the green flag.

That came when my father got home from work and sat down at the table, and nobody took the first bite of food until then. But even he didn't take his first bite until we said grace. Under no circumstances. Momma just didn't allow it.

My momma was an amazing woman. She ran the house and kept us all straight and somehow came out of it with her sanity. If it had been me, all the dirty diapers alone would have been enough to drive me crazy.

With us being stair-stepped the way we were, my momma always had a baby in diapers and usually more than one at a time. Of course, she never had a throw-away diaper in her life, so every diaper she had got washed, hung out on the clothesline and folded.

How the hell did she do it? I sure don't know.

Pat was a big, big help to my mother and Claire maybe not quite as much as Pat but some. From the time Eddie could pick up a wrench, he went to work with my father. Bobby had other stuff he liked to do. He had a couple of friends, and they fished a lot.

If I had to work, I'd work, but I really didn't volunteer for a lot because I wanted to go play or do something else. My younger brother Tommy and I were supposed to cut the grass, so I more or less "elected" Tommy to do most of the yard work.

We were obligated to do household chores and we did them, although I'd have to say it was reluctantly.

Like all kids, we'd put them off if we could or try to get out of them all together. Like all kids, we also used to fight like cats and dogs, at least until the boss lady got involved.

She pretty well ran the show at our house. There's just no other way to say it. She pretty much tried to take care of the little infractions herself. She did everything she could pertaining to disciplining her children until she felt like it had gotten out of her hands, then she called in dad, and that was not fun.

Our biggest fear, all of us, was to hear our mother say, "Wait until your father gets home." That was bad and I don't mean a little bad. That was really bad, because whatever my momma said, my daddy did.

Mr. Mischievous

Some folks who got to know me later in life may find this hard to believe, but at one time, I was an altar boy. Honest, I was an altar boy all the way through the eighth grade.

I would help serve 6 o'clock Mass every Sunday morning at the church we went to. During the week, I would ride a City of Miami transit bus to school and get there in time to help serve two or three masses before going to class.

Of course, being an altar boy didn't stop me from being rambunctious. Honestly, I was one of those kids who was always in trouble, not with the law, but with my parents.

I was one of those kids that if something needed to be done, I'd do it. I was the ringleader. The instigator.

If you needed somebody to get the bicycle out of the driveway without somebody else knowing it, I did it. If you needed somebody to organize the football or baseball game, I did it. If there was mischief to be made, I made it.

Yes, I was the one who would take the initiative and do the stuff I wasn't supposed to do. I was the ringleader on a lot of the stuff we did as kids, but not all of it. Still, I got in trouble for all of it and a lot of times my brothers wouldn't, even though they'd been right there urging me on.

A perfect example was my dad's cigarettes.

"Pop" kept a carton of Chesterfields in the top of his closet, and I was always the one brave enough to go get a pack of cigarettes out of that carton. Then me, Bobby, Eddie and some friends from the neighborhood would go out behind a rock wall close to our house and smoke'em.

Well, when we got caught, they were all just as guilty as me, but I was the one who got my fanny torn up because I got the cigarettes out of my dad's closet.

I got the reputation of being the Allison brother who was always playing, always in trouble and it was a little bit undeserved if you ask me. I liked to play and laugh and cut up, but when I needed to be, I was as serious as anybody. It's just that folks often failed to see that side of me.

Take my swimming and diving. I was very serious about practicing my swimming and diving, but if you asked anybody in my family what I was doing, I was playing at the pool.

I took Tommy to the pool with me one summer, and he ended up getting a job in the concession stand. Suddenly, he was there making money while I was "playing." Suddenly, it was all right for him to go to the pool every day because he was working, but when I went to the pool every day, I was playing. At least that's how a lot of folks saw it.

That reputation stuck with me even as an adult, grown up and married. Everybody thought everything I did was playing. People would tell you that I played all the time.

The truth is, I worked hard to get where I got, but nobody seemed to notice. I worked hard on my swimming and diving and became a champion. I worked hard on my race car driving and won races at the highest level. Shoot, if you look at some of the things I accomplished, I was working pretty hard when I was "playing."

A lot of it was being compared to my brothers. Maybe some of it was because I wasn't the best student in school. Most it of it was probably because people didn't understand how much inner drive and determination I had.

One of my deepest regrets of childhood, though, is that I didn't apply myself more in school. Sadly to say, I was very smart. If I applied myself, I had no problem, but I wouldn't apply myself.

I went to Catholic schools where they were very strict, and I was always in trouble in some way, shape or form with the teachers.

All year long, they would tell me "You're going to fail. You're going to fail." But when it got time for exams, I'd study like the dickens, and I'd pass. It had to be very frustrating for the teachers to see me be able to do it when I wanted to, but not to want to do it all the time.

Finally, I quit school, and that's something else I'm not proud of. I am the only one of the Allison children who lived that did not graduate from high school.

When I quit school after the 10th grade, it was with every intention of going back. I was going to help my dad through a crisis, then I was going to go back. That's really the only reason my mother agreed to let me quit. That and she knew how badly my father needed the help.

I planned to go back, but then I had that motorcycle wreck and stayed in the hospital all that time. That changed everything.

When I had that wreck in March of 1956, I went from a happy-go-lucky kid to someone who had to work for a living. I had bills that had to be paid from my hospitalization and treatment, so I had to keep working.

I'll have to admit I was not in my father's good graces for awhile after that accident, though, because I wasn't at work the day it happened. I remember we had some type of little run-in, and that's why I was at the horse track instead of working that day.

I don't remember what the run-in was about, but my father was probably right, whatever it was. Like most folks, the older I got, the smarter my parents became. But at 15, 16, 17 years old, nobody can tell you anything. Nobody can show you anything.

You can never learn from someone else's bad experiences. You've got to go through them yourself.

I used to tell my kids all the time when they were growing up that if they could learn from just one of my experiences, they'd be ahead of the game. But just like me, most of the time they had to find out for themselves.

One Cool "Pop"

"Pop" Allison was the strongest man I've ever laid eyes on in my life. When we were young kids, he was probably 6-foot, 6-foot-1 and 185 to 190 pounds, and he could do things with his strength that amaze me to this day.

"Pop" installed service station and garage equipment for a living: pumps, tanks, the lifts mechanics used to jack up a car in the bay, that kind of stuff. And we're talking the old-timey ones that were really heavy, not all this new-fangled stuff that's just fiberglass and light-weight sheet metal.

One time, I was working with him to install a lift cylinder in a garage. This was the whole cylinder, the tank and everything in one piece. It probably weighed over a thousand pounds, and part of it had to go in the ground.

We had to dig a big hole, put part of it in the ground and then install the rest of the lift. Normally, we had a big boom truck dad had built himself to help with the lifting. But that day, the boom truck was being used for something else, so the only lifting tool we had was "Pop."

He told me and Eddie he was going to raise one side of the lift so we could slide a 4 x 4 board underneath it and, sure enough, he did it. I don't know how he did it, but he did it.

The lift was lying on the hard ground, and he picked up one end—what he lifted probably weighed about 400 pounds—and held it until Eddie and I slid that 4 x 4 board underneath it.

I saw him numerous times jump in the back of a pickup truck, throw his arms around one of those old-style gas pumps in a bear hug and walk it out the back of the truck.

The pumps these days are just cabinets that have a gauge in them to tell how much gas goes through the nozzle, but those pumps with everything in them probably weighed 300 pounds, and they were taller than he was. I could hardly rock one of them, much less pick one up, but he manhandled them with ease.

I watched him do many a thing like that, and I was always stunned by the man's strength, yet he had the gentlest hands in the world.

When he was working, putting pipe fittings together and such, he knew just how hard to tighten them. Just how much torque to put on a wrench. Just how hard to hit something with a hammer. He was truly amazing.

He was also very mechanical. He loved to work on cars and ride motorcycles, which is probably where we all got our love for the same things. "Pop" was good at that, too.

One of Uncle Jake's favorite stories was about the time when they were kids in New York, and my father built this beautiful motorcycle in the basement of their house. The only problem was he couldn't get it up the steps and had to take it apart again to get it out.

My dad didn't say much. He didn't talk a lot, but he was definitely the coolest dad on the block.

"Pop" was certainly proud after another of my successful races at a track in South Florida. (**Courtesy of Al Arias**)

As I said, he'd come out and play ball, no matter how tired he was from work. He liked to go to Lake Okeechobee duck hunting with Uncle Jake, and as we got older, he'd take us, too.

He used to take the whole family to Matheson Hammock Park there in Miami. We had a little wooden boat, and he'd take us boys out fishing in it. We'd catch sand perch, then we'd come in and clean them, and the whole family would have a fish fry. We might do that twice a year.

He'd take us places and do things with us when he could, but he worked the rest of the time. That's another thing I'll always remember about "Pop" Allison. He was one of the hardest-working men I've ever known.

I guess he had to be. He supported a wife and a family and kept clothes on our backs and food on our table. He put 10 kids through Catholic school. He never let his family want for anything we needed.

One other thing I'll always remember about "Pop" is how much he loved us. He may not have been a big talker, but we definitely knew he loved us. And buddy you didn't mess with his wife or his kids. He'd take your head off if you did.

He'd tear our butts up, too, especially if momma said we needed it. If she said one of us needed a spanking, he did it, and if he got after you, you remembered it. The thing that struck me, though, was that he always seemed to stop three or four licks before we got what we probably deserved for our troubles.

Maybe he was really a softie at heart when it came to punishing the children he loved so much. Or maybe he just knew his own strength.

We lost "Pop" in April of 1992 at the age of 86, and I miss him to this day.

I owe "Pop" all the thanks in the world for working so hard to take care of all us kids, to raise us right and make us what we turned out to be today. I also owe "Pop" thanks for something he didn't do.

He never once said anything derogatory about racing. Maybe it was because he loved working on cars and riding fast

"Pop," Mom, and I celebrate in Victory Lane at Daytona on July 4, 1970, after I won the Firecracker 400. (**Motorsports Images & Archives Photography. Used with permission.**)

motorcycles and such, but I never once remember him trying to discourage any of us from racing.

My mother, on the other hand, thought we were all going straight to hell when we all got the racing bug. She just didn't think it was an atmosphere we should get into, but to her credit, she never outright forbid us to do it, either.

It's a good thing, too, because if she had been more vocal about it, my dad might have been more inclined to put a stop to it. After all, what momma said, daddy did.

Or maybe he would have just said to himself, "They're going to do it anyway," because we were, and went on. Like I said, my dad was cool.

How It All Started

As a kid, I preferred roller skating to racing any day. I roller skated every Wednesday and Saturday night, and I loved it. In fact, I cared so little about racing somebody had to bribe me to get me to my first race.

A guy who lived four houses down from me in 1958—I don't even remember his name—came down one day and said he had a racecar and wanted me to go to the track with him the next Saturday night. I told him I wasn't going to no racetrack, I was going roller skating.

He said he really wanted me to go and he'd pay my way. He kept on until I finally said I'd go, but I didn't plan on making it a habit. Boy, that sounds funny to say now, considering how many racetracks I've been to in my life.

We went to Hialeah Speedway, and there must have been 60 cars in the pits that night. He had what they called an amateur car with a two-barrel carburetor, not a very safely built car, but at the time we didn't think about that.

His practice time came, and he went out. Then he came in the pits and said, "How'd I do?" I told him I could drive that car better than that, and I'd never driven a racecar in my life.

He didn't get mad, though. He just told me to go run the car in the next practice. I told him I'd never driven before, but still he told me I probably could drive the car better than him, and I was going to race it that night. So I did.

I started in the back and finished in the back, but that didn't matter. All that mattered was I had gotten my first taste of racing, and it was sweet.

We went home that night, but instead of stopping at his house, he pulled into this big driveway that went around behind my house. Then he got out and started to unhook the tow bar from his car to leave the racecar in our yard. I told him, "Man, you can't leave this here. My mother and father will call the cops and have this thing hauled off."

He said he was letting me have the racecar. I told him I didn't want it, but he told me to take it and run it because I had done a much better job in it than he had, and he was right. What's weird about the whole deal, though, is that I don't remember ever seeing the guy again. It was like he gave me that racecar then disappeared off the face of the earth.

So that's how I got may start in racing. One week later came the incident that caused it to consume me.

I went back to Hialeah the next Saturday night and ran pretty good. The next day, there was a race scheduled at Pal-metto Speedway, too, and Bobby had been outrunning them there pretty regular, so they were going to protest his engine after the race.

To do that, the other drivers had to post a $75 protest fee and, if Bobby was found to be legal, he'd get the $75. He had gotten wind of the protest, and he was legal, so he was already counting on $75 going in his pocket.

In my heat race, I broke an axle key, and on the old model Ford I had, you had to pull the whole rear end apart and take the axle out to fix it. I didn't have the tools I needed to fix it, so I was done for the day.

Bobby won the heat race and a match race, and they were getting ready to have a consolation race. He was entered in that, too, but somebody somehow talked him into letting me drive his car in that race. He didn't want to do it, but he agreed.

So there I was out there in his car, and it was only my second weekend in organized racing. Of course, that never crossed my mind at the time. Instead, in typical Allison fashion, I was out there saying to myself, "If he can win races in this car, I can, too."

I proceeded to drive the car straight into the guard rail, which at Palmetto was made out of railroad track iron. It destroyed Bobby's racecar, and I hit my head and put a big dent in the little old half helmets we wore at the time.

They hauled me to the hospital, and Bobby hauled the car back to my daddy's shop, steam blowing the whole way, I'm sure. They checked me out, found I hadn't done any major damage and released me.

I left the hospital and went to the shop where Bobby was already cutting that car up. He took the motor out and the seats, what good stuff he could get from it to build a new racecar. I walked in and started to go up and say something about being sorry for wrecking his car when he turned around and ripped into me.

Here I'm taking the checkered flag at Medley Speedway in a pink car called "Little Joe" that Bobby had wrecked previously, and I had put back together. (**Courtesy of Bobby 5x5 Day**)

He was already mad, and when he saw me walk in, it made him even madder. I guess he figured since I wasn't laid up in the hospital or dead, I was in good enough shape to take a tongue lashing.

He lit into me and said he couldn't believe he'd let them talk him into letting me drive his car. He said I couldn't drive a lick. Then he said something that has stuck with me since. Bobby Allison looked straight at me and said, "You'll never make a racecar driver."

At that moment, my main mission in life became to prove him wrong.

I don't remember the entire text of our conversation that day, but I've never forgotten that statement. That was a direct quote, and I remember it just like it happened yesterday.

For my own brother to say that put a thorn in my side that never went away. He did it out of a fit of temper, sure, but once he said it, that's where we stood. That lit the fire in me to be the best racecar driver I possibly could.

That was the way I was my whole life, with my diving, my pitching, my roller skating, my racecar driving, anything. If you told me I couldn't do something, well, I would show you.

Looking back, I believe what made Bobby so mad wasn't so much that his car got torn up, but that he didn't get that $75 of protest money. That was a lot of money in 1958, and he wanted that money so bad it wasn't funny. He was going to get it, too, because he was definitely legal.

What he got instead was one of the toughest rivals he'd ever have to race against in his career.

This was the 1933 Pontiac Sedan that I dominated with at Dixie Speedway in Midfield, Alabama, and numerous other short tracks around the southeast in 1962. **(Courtesy of Donnie and Pat Allison)**

"Straighten It Out"

I love my brother very much, and always will. But when we buckled up the seatbelts and chin straps in a racecar, you couldn't find two fiercer competitors.

When we were in our racecars, he was my brother, but he was also another racecar driver, and I wanted to beat him as bad—or worse—than anybody else I raced against. So I drove especially hard when I was racing Bobby.

I don't mean I ran over him because I never ran over anybody. But knock him around? If he did something I didn't like, yes sir. At that point, it didn't matter that he was my brother.

If he put something on me, I put it back. I did that with everybody I raced. If they didn't dish it out, I didn't dish it out. If they dished it out, I dished it back, and I did it right then. I didn't mess around.

Bobby, Eddie, and I were smiling in the pits before this race, but when it began you can bet those smiles went away because we all took our racing so seriously. **(Courtesy of Donnie and Pat Allison)**

Bobby and I have both always been very competitive, head-strong—some would probably say stubborn—people who have never minded saying exactly what we think to anybody, especially each other.

I don't think Bobby has ever had anything he wanted to say to me that he didn't say, and I know I've never held anything back from him, either. If I wanted to tell him something, I told him whether he liked it or not. It didn't make any difference to me.

Considering all that, it's no wonder Bobby and I have had our differences over the years. Only one time did we have any problems on the track, though.

That came in 1963 when we were both still tearing it up in our short-track cars all over the Southeast.

Bobby and I were family off the track and the fiercest of competitors on it. I wanted to beat him as much or more than any other driver out there, and I'm sure he felt the same way about me. (**Courtesy of Donnie and Pat Allison**)

One Friday night early in the season, I wrecked my car during a race at Birmingham International Raceway, a 5/8th-mile track in Birmingham, Ala., that we considered home. At the time

AS I RECALL...

I was working with my brother, Eddie, and we were supporting my family and his off what I won with that racecar.

The car was a mess, and we didn't have the money to fix it, so when I got an offer from a man named Charles Morgan to drive his car, I told Eddie I was going to take it. It sounded like a good idea to him until we could get enough money to fix our car, so I went to race for Charles.

The next Saturday night I went to Montgomery Speedway in Montgomery, Ala., and I was the fastest car there. Bobby had raced at Nashville the previous week and won, so they said he had to start in the back, and I had a new car and had to start back there, too.

Well, that hacked Bobby off, so he was steaming when we went into turn three on the first lap of the heat race. We went in, he spun me out and I stuck Charles' car in the wall backwards and tore up the axle. It messed up some other stuff, too, so we couldn't race the car anymore that night.

I was furious. There I was trying to support my family and Eddie's, too, and I had wrecked two cars in two weekends. I was so mad I walked out on the track as Bobby came by on the caution lap and shook my finger at him.

After the race, I walked over to Bobby and said, "I'll wreck you for that." That's all I said and, to be honest, I don't remember what he said back to me because I was so mad I didn't listen.

Eddie and I loaded up and headed to Birmingham after the race and went to work. We worked all night and got the car ready for a 100-lapper the next afternoon at BIR. We put in a new axle and cross member and had it ready to go.

The car ran good and, as the race wound down, I was leading, and Bobby was second. I remember thinking that was a whole lot better than wrecking him because he couldn't stand to run second to anybody, especially me.

On lap 92, the caution came out. On the re-start, my car stumbled for some reason, and he passed me going into turn one. So I had to decide if I was going to keep my word or eat it.

As the laps counted down, I kept saying, "You said you were going to wreck him. You said you were going to wreck him." Finally, on the white-flag lap, I made up my mind.

We went into turn one, and I put the bumper to him. He spun 360 degrees, I shot by to win the race, and he still finished second.

After the race, I pulled up to the start/finish line and got out to get the flag and my trophy. As I did, Bobby came along real slow and shook his finger at me. So I gave him the No. 1 sign right there in front of the grandstand.

Just then, I felt a big hand come down on my shoulder and heard a voice say, "I didn't raise you boys like that." It was my dad, and he was hot.

He told me he wanted me to go to Bobby and straighten it out. I told him I didn't start it, so I wasn't going to be the one to go to him and straighten it out. He said, "You know your brother will never come to you, so I'm telling you to go to him and straighten it out."

When "Pop" told us to do something like that, we did it. So the next day I went to see Bobby.

A guy who owned a car lot in downtown Birmingham sponsored Bobby at that time, and that's where I found him. I pulled up in the alleyway behind the office and got out of my truck. As I did, one of the salesmen came out the door and saw me, and his eyes got as big as saucers.

I asked if Bobby was there, and the salesman told me he was inside, and then moved aside to give me plenty of room. I went in, and Bobby was in the office with two other salesmen. I looked him straight in the eye and told him I was there because my dad had told me to come get things straightened out.

I told him, "We can finish this here or in the alleyway. It doesn't make any difference to me, but we are going to get this behind us today."

We had a pretty stout discussion you might say, and I told him I was going to tell him my side of it, and he was going to listen whether he liked it or not.

I told him that the way things had been going, if he was in the front and I was in the back and we had a problem, he claimed it was my fault. Or if I was in the front and he was in the back and we had a problem, he claimed it was my fault. Or if I was on the right and he was on the left ... well you get the picture.

I asked him if he had ever stopped to think that 50 percent of the time he was in the same place I had been when we had the problem. Then I pointed out that if half the time he was where I had been, it couldn't be my fault 100 percent of the time.

He admitted he'd never thought of it like that, and we never had another run-in on the track again.

All it took was me explaining it forcefully enough for him to finally get it through his hard head.

Bobby's Way ... Or the Highway

There have been many times in our lives when Bobby has had an opinion on something that has been totally different from mine. Actually, most of the time, Bobby's opinion has been different from mine.

Aside from our differing views on my future racing prospects, there were plenty of other things we didn't see eye-to-eye on. One I remember most involved his son, Davey, and Davey's first racecar.

When Davey was trying to get started in racing in the early 1980s, I didn't feel like Bobby helped him in the right way. Bobby had his ideas on how to help Davey, but I watched how hard the kid worked, and it upset me because I didn't think Bobby was doing enough.

I finally went over to his shop in Hueytown one day and sat there in his office and told him, "Bobby, help the kid get a racecar; he's ready." Bobby said no, it was fine just like it was, and I told him Davey was ready for a racecar.

The discussion got pretty heated, and I got a little aggravated, so I went downstairs and hollered for Davey to hook his trailer to his truck and come over to my shop. I had decided to give him an old Nova I had that had been a pretty good racecar in years past.

Well Davey almost fell over. Then he hooked that trailer up so fast he almost beat me to the shop.

I gave him that racecar, and he took it and worked on it. He ran it that Friday night and won a race. In fact, he won two races in it that first weekend. It was ironic because I had been the one saying he needed a racecar so he could win races. Then I gave him one, and he won the first two times out and made me look pretty smart.

The next Monday, Bobby called me and asked me to come to his shop. I went over, and when I walked in the office, he said, "I guess you were right." I told him it wasn't a question of being right or wrong there. It was just a case of seeing what needed to be seen.

We disagreed on our approaches to racing, too, and we had many a conversation about it.

I didn't go somewhere just to be out there riding around. If I didn't think I could win, or at least be competitive enough to have a shot at winning, I didn't go. Bobby wasn't that way. He would go race anywhere, especially if they paid him any kind of show money.

After I started winning races consistently in my short-track car, I got offers to go places like Wisconsin to run against the ASA guys. I had a good car where I was, and I knew I probably wasn't going to go up there and beat them on their home turf, so I didn't go. And I was satisfied with that.

But it gave me a reputation that I didn't want to race that much, one Bobby even helped along the way when people would ask him about it. The truth wasn't that I didn't want to race. It was that if I thought I was going some place where I was just going to ride around in circles, I didn't see any sense in making the trip.

One other story comes to mind when I think about how headstrong we were and how much we butted heads in our younger days.

In the late 1970s, Neil Bonnett was driving a short-track car Bobby owned. A guy named Charlie Wright was the crew chief, and they were doing well.

Neil was getting ready to go race in Nashville one weekend, and I had just raced there the week before and really tuned their guitars for them. So Neil asked me what I did differently on my car that worked so well.

I told him some changes to make in the front end and some things to do with the wheels. So he went to Nashville and won the race. He came back and told me he couldn't believe how good that car ran with those changes.

I asked him if he told Bobby what he'd done, and his eyes got wide. He said, "No, no, no. We put it back like it was. If Bobby Allison found out we did that, he would fire me." And Bobby would have, just because I had told them what to do, and it worked.

No, you were supposed to do what Bobby said do in those days. But I don't think he did it just to be top dog or anything like that. He was just that firm a believer in whatever he said and that confident in his own abilities.

I saw early on, though, there was more than one way to skin a cat. The main objective was to get the skin off, and how you did it was your business.

Fair and Square

One thing I will say about Bobby Allison is that, regardless of whether our personal relationship was strained or not at the time, I never once worried about him doing something to put me in danger on the racetrack.

I knew he would take advantage of me if he could, just like I would take advantage of him if I could. But I never in my life felt like Bobby took a cheap shot at me, and we had some great races over the years because of it.

The first time I beat him was in 1960 in a modified race. I don't even remember where the race was run or where Bobby finished. All I remember is how satisfying it was to show people—especially Bobby—that I *could* drive a racecar.

It was particularly satisfying to win when we both had good cars and were up front running one-two for the lead. A lot of people may not remember it, but it happened five times in Grand National or Winston Cup races, and I won four of those.

My very first Grand National win came at North Carolina Speedway in Rockingham in 1968, and Bobby ran second. He was over two laps behind when we finished, but he ran second.

It happened again in the 1969 National 500 at Charlotte, the 1970 Southeastern 500 at Bristol and the 1971 Winston 500 at Talladega. In all three races, Bobby was again the driver with the next best view of my rear spoiler as I took the checkered flag.

He finally got me at Charlotte in the 1971 World 600, but by that time I had written something in the record books that can never be changed. Out of the 25 times brothers have finished

Five times in our careers Bobby and I finished one-two in a race and four times I was leading at the checkered flag. (**Courtesy of Donnie and Pat Allison**)

one-two in all of NASCAR history—and four of the five times we did it—it was Donnie Allison in front.

We've never talked about it much, but I know at the time that really stuck in his craw. He loved me as much as any brother could, but I can tell you nothing made him any unhappier than losing like that to me. My satisfaction came from the fact that he could never change it.

Something else that won't change, at least not anytime soon, is the fact that the Allison brothers have won more races as a pair than any other brother combination in NASCAR history.

Bobby has 84 career wins, though by all rights he should have 85 since he won a race at Bowman-Gray Stadium in 1971 that NASCAR refuses to give him credit for. He drove a Mustang owned by Melvin Joseph—the same one he won Sportsman Division and Winston Cup races at Talladega in that same year—to victory that night.

Granted, it was a Grand American Series car with a smaller engine and tires, but NASCAR had relaxed the rules at the time because it needed the Grand Am cars to fill out some of the short-track fields. Still, for reasons they have never been able to explain to Bobby's satisfaction, they don't list it in his career victory column.

So his 84 victories tie Bobby for fourth on NASCAR's all-time list with Darrell Waltrip, and he won 58 poles. Together we have 94 career victories, six more than D.W. and Michael Waltrip's 88.

I'm very proud of what we accomplished together as racecar drivers. How we left our mark on the sport. How we made the Allison name synonymous with winning.

I must admit something here, though. Maybe it was because Bobby pissed me off so bad when he said I'd never make a racecar

DONNIE ALLISON

driver, but for a long time I resented having my career so closely measured by the yardstick of his success.

I felt like for a long time I had to look out from behind his shoulder to get people to notice me, even though I earned my racing stripes pretty early on in my career. It was the big-brother thing, I guess. He was older and started first, and he was so successful it was inevitable people were going to judge me by him.

Folks went overboard with it sometimes, though, and that rankled me.

For example, in 1962 Eddie and I were having a pretty rough time of it because we had a Buick engine in our short-track car, and we just couldn't get it to run. Bobby had a 327 Chevy engine in his basement, and I went to him and asked him if he'd sell it to me.

He said he'd sell it for $600. I told him I didn't have the money to pay him up front, but I'd get it as soon as I could, and he told me that was fine, take the engine. I borrowed a set of injectors, tuned it up, and in two weeks I had paid for everything.

I won the mid-season championship race at Dixie Speedway in Midfield, Ala., and by the end of the year, I had won 36 races.

When that season was over, Eddie and I had done all the work, but Bobby still got some of the credit because he sold me the engine. A lot of people said, "Well your brother sold you a good motor," which was fine because he did, and I appreciated it. But he didn't fine tune it all week. He didn't drive the car every night.

I think back on it, and there were a lot of times in our careers when something happened, something I did, and people thought it was because he did something for me instead of me doing it for myself.

Shoot, Bobby was even the story the day I got my first Grand National win.

I had just gotten the biggest win in my life and put my car owner, Banjo Matthews, back in victory lane for the first time in years, and all the press wanted to talk about was Bobby. The media made a big deal because Bobby finished so high that day driving a Chevy for J.D. Bracken—an owner not associated with any of the big-money factory teams—after he'd quit the Ford factory team earlier that year.

It was only natural, I guess. I was the younger brother doing the same thing he was. Bobby had the ball first and, all of a sudden, I came along and started trying to take it from him. It had to have an impact on our relationship.

The best example I can think of today is Peyton Manning and Eli Manning. They are both great quarterbacks who have been and will be compared all their lives. So who's the better quarterback? I'll bet if you ask Peyton, he'd say Peyton, and if you asked Eli, he'd say Eli.

So who was the better racecar driver between the Allison brothers? If you ask Bobby, he'll say he was. If you ask me, I'll say I was without any doubt in my mind.

If he wants to disagree, fine. It won't be the first time we've done that. But it still won't ever change my mind.

Setting the Record Straight

Over the years, the way I remember things about our childhood and our careers and the way Bobby remembers them have often been entirely different. One story in particular I want to clear up involves our leaving Florida to come to Alabama in 1959.

The way Bobby remembers it, he went to our mother and told her he was going to Alabama with two guys named Kenny Andrews and Gil Hearn. They told Bobby they were going back to Alabama because they'd been there, and the racing was great. Then they asked if he wanted to come, too.

Bobby remembers my mother telling him that was probably what he should do and that he should take me along. He remembers her saying, "I'd appreciate it if you'd take your brother with you because he's driving your dad nuts."

He'll tell you he knows I hate for him to say that, and he's right, because as far as I know, it's not true.

I had never in my life heard that story until it came out in Bobby's book a few years back, and that was a long, long time after we left Florida for Alabama. Besides, I asked my mother about it long before she died on March 6, 2008, at the age of 101, and she didn't remember saying it, and you can't tell me she wouldn't remember a conversation like that.

Here's what really happened:

Bobby came to me and said he was going to Alabama with Kenny and Gil and asked if I wanted to go, too. I said I did, but I didn't have the money to go. He said I'd have to get some money from somewhere, so I hocked a Winchester 16-gauge pump shotgun that I wouldn't have traded for the world otherwise.

I had saved every dime of the money to buy it, and then I hocked it for $35 to a man that owned a junkyard next to my dad's place. He promised I could buy the shotgun back when I got home, but later, he said he'd given it to his son, and I couldn't have it.

I was so mad I couldn't see straight because I wanted that shotgun back, but I never got it. So I lost my prized shotgun in the deal to go to Alabama and help Bobby get started racing there. But I eventually gained a ride in a racecar, and I think it was worth it.

The point, though, is that if my mom had truly wanted Bobby to take me with him because I was driving my dad nuts, she could've said "Here's $20, take Donnie with you." It was $15 less than I got, but I would have had some money, I would have been out of their hair, and I would have still had my shotgun. She never said that, though, because it never happened.

No, I don't believe my mother ever asked Bobby to take me away. I know I would not have hocked a shotgun I loved so much if coming to Alabama had not been totally my decision. Also, I'm sure if my dad had a problem with me, he wouldn't have chosen to deal with it by sending me out of state.

That's because I remember how "Pop" Allison was, and even though I might have been 20 years old at the time, if I had been driving him crazy, he would have dealt with it in a much more direct fashion.

You can be sure if there had been a problem, we would have taken a trip out behind the junkyard, and it would have come to a stop right there. After that, I wouldn't have felt like driving anybody else crazy.

Even though it may not have looked that way at times, Bobby and I have always been close. Sure, we've cussed and fussed and bickered with one another over the years, but that's what brothers do.

It's a family thing. We might have fought like cats and dogs among ourselves, but nobody else better mess with us. Otherwise, you'd have both of us to deal with, just like Cale found out in Daytona.

We've looked out for one another in other ways, too.

When we went to Alabama, I was the crew, and I worked hard to help Bobby get going there. We both worked hard. I'd say the two of us worked harder in our first two or three years in Alabama than a lot of men worked in a lifetime. It was tough, but we loved every minute of it.

I'll never forget those days. Money was so tight there were times we'd buy a bushel basket of peaches for $1, and we'd eat peaches for breakfast, lunch and dinner until the basket was empty.

And if it wasn't peaches, it was pancakes. We ate a lot of pancakes for breakfast because you could get two of them for 29 cents. No meat and no eggs because that was 39 cents extra, and sometimes that stretched our budget too thin.

For lunch or dinner, we'd usually eat fried chicken because in most places we went in the South, you could get half a chicken for $1.25. Every once in awhile, we'd treat ourselves to a hamburger steak for dinner and, every now and then if one of us had won a race, we'd get a real steak.

As we started winning, that happened more often, and by the time we got to Grand National racing, we were eating real steak pretty regularly.

We've helped each other take care of our families and keep our racing operations going, too.

When Bobby started Bobby Allison Racing back in the 1960s, that was his and Donnie Allison's money. We bought the property together in Hueytown where Bobby has a house and shop to this day.

We divided that property up, a lot for him and Judy and a lot for me and Pat, and built houses on them. Then we built a racing shop together.

At that time I was driving for Banjo Matthews and doing pretty good. So I paid the money out of my bank account to build the original shop there because he just didn't have it. I had no qualms about that. It's just the way we've always done things.

When he was doing well and I wasn't, if I needed something, all I had to do was tell him and I got it, like that motor back in '62. If I was doing well and he wasn't, all he had to do was say something to me, and he got whatever help I could give.

It didn't make any difference if it was a part for a racecar or cash out of our pockets, we shared it. That's the way we were and are to this day.

Bobby was stubborn, hard-headed, bossy and could make me so mad I'd want to strangle him sometimes, and he'd probably say the same about me. But I don't think either of us would trade those times now if we could. I know I wouldn't.

Racecar drivers, and men in general, often aren't good at expressing their emotions, especially to other men. But I think Bobby knows I love him, and I know he loves me.

How do I know? Well, it goes back to our racing careers. I honestly believe that if Bobby ever gave anybody a break on a racetrack, it was me. I can't think of a specific instance, but I feel like it happened somewhere along the way, and that's very telling because Bobby Allison just didn't give people breaks on the racetrack.

If it happened, even once, in all the races we ran against one another, it was because we were family. And family takes care of family.

She may not have liked me going racing at first, but my mother—shown here giving me a big hug in Victory Lane after I won one of the Twin 125 qualifying races in Daytona in 1980—became one of my biggest fans. (**Courtesy of Donnie and Pat Allison**)

The Alabama Gang

When Bobby and I were just getting started in south Florida in the late 1950s, there was a guy there named Charles Farmer who was already a hotshot. He had red, wavy hair, so naturally everybody called him "Red" and man could he drive.

I loved to watch "Red" Farmer drive a racecar. He had good cars, and if he wasn't winning, he was running second or third. Wherever he placed, he was plenty of fun to watch.

We were only acquaintances at that time. Bobby had done a little bit to help "Red" out, but I didn't know him that well. That changed a few years later when we all ended up together in Alabama, where we began to run and take the money.

"Red" had heard the stories about how good the racing was in Alabama, too, and he decided to find out for himself. He loaded up his mechanics, Wayne Kackley and Homer Warren, and his 36 Chevy Coupe and began rolling towards Bama like a tide.

I was racing at Dixie Speedway at Midfield, Ala., at the time, and I got word that "Red" was coming. I started telling the guys there, "Look out, 'Red' Farmer is coming, 'Red' Farmer is coming," and they just laughed.

They laughed even harder when "Red" pulled in towing his car on a trailer about three feet off the ground with a station wagon. They went into hysterics when "Red" put his car in gear and backed it off the trailer.

Everybody in Alabama in those days ran straight drives in their car—meaning they had no reverse—so you had to push the car off the trailer because you couldn't shift gears. "Red" also had two four-barrel carburetors under his hood when all the Alabama drivers were using injectors. So they thought he would be easy pickings.

I told all those guys they wouldn't be laughing at the end of the night, and sure enough, they weren't. "Red" won every race that night at Dixie Speedway, then went to Montgomery Speedway the next night and won all the races there. At that point, he decided the racing was pretty good in Alabama and that he'd stay, too.

So the Alabama Gang was born.

When Bobby, "Red" and I got together we became fast friends and one big racing family, and the other guys never really had a chance. Over the next few years, we dominated short-track racing in the Southeast to the point other drivers hated seeing us pull into the pits as much a bank teller hated seeing the James boys come through the door. They hated seeing us because they knew we were going to take their money, too.

I never gave it any thought at the time, but looking back, that had to be hard on those guys.

In 1962, we ran 106 races at tracks all over the Southeast—Birmingham; Montgomery; Nashville; Columbus

"Red" Farmer, Bobby and I made up the the famed (and feared) Alabama Gang. Here we are with our rides on display before a race at Bristol in 1961. (**Courtesy of Donnie and Pat Allison**)

Ga.; Chattanooga—and between us we won 96. That means out of all the drivers in all those races, only 10 of them got to experience the thrill of winning—and collect the paycheck that went with it. But at the time, we didn't think about it. We were just trying to figure out why we didn't win those other 10.

The Alabama Gang officially got its name from a man named Bob Harmon, who owned Montgomery Speedway.

He used to say that after we started winning everything, all he heard people talking about was, "the Alabama Gang, the Alabama Gang." So he'd get on the microphone at the track, hours before the races even started as the fans were just filing in, and talk about, "the Alabama Gang is coming," or "the Alabama Gang is here," and it stuck.

Bobby was the leader of the Alabama Gang. We all won a lot of races, but for a few years there Bobby won the most, so that made him the leader. Plus, he just kind of assumed the role because everybody did what Bobby said do anyway.

A perfect example was when we built "Red" a whole new racecar.

Not too long after he got to Alabama, "Red" crashed in Nashville and totaled the car he'd brought with him from Florida. So Bobby came to me and said he was going help build "Red" a new racecar, which meant he expected me to help, too.

I told him I wasn't sure that was a good idea, but he said we were going to do it. We both had gorgeous new racecars, sedans that were really fast, but Bobby had confidence that, in equal racecars, he would come out all right.

I didn't feel that way. I was just starting. I was struggling at the time. I didn't want to take unfair advantage of somebody, but I didn't want to help them beat me, either.

Well, we built "Red" a new racecar, and the first weekend he had it, he ran three races and won all three. I asked Bobby, "Now you got any reservations about building 'Red' a new racecar?" But he said we'd be all right, and he was right.

By the end of the 1962 season, I think Bobby had won 36 races, I had won 33 and "Red" had won 29 or 30.

By 1963, Eddie had also made the move to Alabama, and he and I were dusting everybody good. We had a modified special with a 471 GMC blower, and it was so good that the next season they changed the rules so folks couldn't run it anymore.

That year, "Red" was going to run a Ford sponsored by a Birmingham car dealership, and he just knew he was going to win all the races. I had moved to Chattanooga, Tenn., to drive for a man named Bob White, and a guy named Bob Wright was my crew chief.

We built a 58 Ford that was fast and went to BIR for the first race of the season and won. Then we went to Montgomery for the first race of the season and won it, too.

Now this hacked "Red" off for some reason and fanned the flames on a pretty good spat between us.

"Red" really got on the gas after that, and I don't think I won another race at BIR all season. But after winning the first one at Montgomery, I missed out on the next one, then won five in a row. And the only reason I didn't win all of them was because Red and I crashed one another coming off turn two.

For some reason, a little friction had been building between us for awhile. I don't really blame it on "Red." I think his crew aggravated him all the time about me for whatever reason. But whatever it was, it had started to spread from the crews to the families, and that was where it had to stop. It came to a head that night in Montgomery in a pretty ugly way.

Late in the race, we got together and the collision broke the valve stem out of his left-rear tire. It went flat, and he spun out. It made him so mad he went in the pits, changed the tire and came back on the racetrack to wreck me and missed.

I was mad, but I was leading the race and still wanted to win so I drove on, but "Red" tried to wreck me again on the final lap, and I ran him up and over the bank coming out of turn two. At that time, Montgomery Speedway didn't have a concrete wall or even a guard rail, and I thought about following him over the bank and trying to run over him. But I was still in the racetrack, so I straightened myself out and ran second to Friday Hassler.

I will admit, though, I was so mad at "Red" right then that if he had come back over the bank and I had been pointed at him, I would have driven straight through him as hard as I could.

I didn't say anything to "Red" after the race, but I told Bob Wright it was going to be bad before it got over because I wasn't going to take it anymore.

The next day we went to BIR, and Johnny Morrison, the pit steward there, was also the pit steward at Montgomery Speedway so he knew what had happened the night before. And being the typical NASCAR inspector of the day who liked to show his authority, Morrison called us together for a chat as soon as we got there.

He told us he didn't care what had happened at Montgomery, he wasn't going to stand for any wrecking at BIR. He said it was over with and that we had better get together and straighten it out if we wanted to race that day.

I looked "Red" straight in the eye and said, "We can do one of three things—we can race, we can wreck or we can fight, and it doesn't make any difference to me which one we do." But I told him if he wanted to wreck he "better show up at 4 o'clock in the afternoon because when you unload your car, I'm going to drive right through it even if it's sitting in the pits."

He looked me straight in the eye and said, "It's all behind us," and we never had another problem on the track.

Considering the close competitors we were, we all got along exceptionally well. Aside from that one run-in, "Red" and I never had any problems, and Bobby and "Red," to my knowledge, never had any at all.

It's a good thing, too, because we were always together.

We traveled together. We raced together. Heck, from the time we all got to Alabama until 1962 when "Red" bought the house in Hueytown he lives in to this day, we all shared an apartment in Fairfield, Ala.

I grew to love "Red" like a brother. An older brother. A much older brother.

After spending so much time with him over the years, I've earned the right to give "Red" grief about his age, and I do, as much as possible.

Over the years, "Red" has refused to tell people how old he is, claiming he doesn't know for sure. People used to ask me about it, even back then, and I'd tell them I didn't know, but he was already drawing a Social Security check.

I thought his wife, Joan, had let me in on the Farmer family secret after I got hurt in Charlotte in 1981. She told me then she was 53, and she was one year older than "Red." But the truth is, I'm just like everybody else. I've known him for about a hundred years and still have no idea how old he really is.

But he's still going strong in a racecar.

"Red" still runs a regular schedule on a dirt track just down the road from the Talladega Superspeedway. He'll tell you he does it because racing is all he knows how to do, but my guess is, it's also because he still likes showing those young guys a thing or two.

"Red" Farmer won consistently at every level he ever raced at, but he never made the move full-time to Grand National or Winston Cup. He ran some races on those circuits, but he never made the move permanently because, I think, even back then the politics involved in big-time racing didn't sit well with him.

He was also a realist. He knew a driver didn't have much chance to win if he didn't drive for one of the big factory teams, and the factories weren't looking for drivers with as much mileage on them as "Red" had.

So he was content to stay where he was comfortable and where he was winning races by the hundreds. But let there be no mistake about it. If "Red" Farmer had gone to Grand National or Winston Cup full-time with the same mindset, determination and skill with which he drove his other cars, there's no doubt he would have been a champion.

I get asked a lot, which drivers were members of the Alabama Gang? The answer is simple. The Alabama Gang was, and is, Bobby Allison, "Red" Farmer and Donnie Allison. No more. No less.

As it evolved, other Allison family members and other Alabama drivers were associated with the Alabama Gang, but that was just the media's doing. To me, the Alabama Gang was me, Bobby, "Red" and Bob Harmon because he was the one that first promoted it and made it famous.

Those that came along later were not members, to my way of thinking. Not Neil Bonnett. Not Davey or Clifford Allison. Not Hut Stricklin. Not Stanley Smith or Mickey Gibbs. They weren't members simply because of circumstances.

We worked hard in those days to earn our reputation, and I felt like the title of Alabama Gang was an honor bestowed on us by Bob Harmon that recognized our hard work.

Those guys that came later didn't travel the roads and turn the wrenches and win all those races with us in the early years, so they couldn't be Alabama Gang members. So for some sportswriter pushing keys on a typewriter or a computer to put Neil or

Davey or Hut or anybody in the Alabama Gang by association was just not within their right to do.

I love all those guys like they were my own and am proud to have been associated with them as racecar drivers over the years, but it still burns me a little bit when I hear all those names mentioned because the Alabama Gang is a pretty exclusive club. And only Bobby, "Red" and I are lifetime members.

Pat and the Kids

Before I talk anymore about my racing, I have to talk about someone who has stood behind me nearly every mile of the way despite a rather damp start to our relationship.

Let's just say when it came to finding the woman of my dreams, I almost ruined everything by throwing cold water on it. Twice.

I was a 20-year-old free spirit playing tackle football with a bunch of boys at a park near my parents' house one Sunday afternoon when four girls stopped by to watch. One of them was named Pat, and I had never seen her before, but she caught my eye.

There was a little canal that ran right next to the park that we swam in sometimes, and after the game, somebody got the bright idea to throw those girls in. So I threw Pat in, and I'm sure it upset her, but to her credit, she didn't make a big deal about it.

Her friend, Janet Raether, came over and told me her name was Pat Leserra and asked if I would take her home to change clothes since I was one of the few guys who owned a car. In typical macho fashion, my response was, "I didn't bring her, I'm not taking her home."

They got a ride home with somebody else, but Pat was a looker, and she had my attention. After they left, my friend Bobby Hammond came over and asked me if I wanted to go out with those girls that night. I said, "Only if I go with Pat Leserra."

We all went it out, but it wasn't really a date. We just went to a drive-in restaurant, then I took her home.

A couple of days later, Bobby came over and told me that Pat was coming over to a girl's house up the street where there was a pool and that we should go swimming. I went over and Pat was standing there in a big wide skirt like the girls wore then and a white blouse with big straps over her shoulders.

She was only 15 years old at the time, but I had decided she was the most beautiful thing I had ever seen. Of course, my way of expressing that to her was to throw her in the water ... again. Then I refused to take her home ... again.

I found out later Pat had to ride a City of Miami transit bus all the way home in those wet clothes, which had I known might have a made a difference in me taking her home. Or it might not have.

At that time, I had kind of sworn off having a girlfriend. In my middle teens I had one I would have done anything in the world for, but she did me wrong—at least by my way of thinking —and spoiled me on all the rest.

It was silly, but her injustice was not being ready to go when I was ready to come by and get her.

I had a job in downtown Miami, and I called her one Friday and told her I was going to have to work late. She said that was

OK, just call her when I got home and she'd be ready to go. Well, I got off a good bit earlier than I expected, went home, got changed then called her and told her to be ready in five minutes.

She told me it would take her more than five minutes to wash her face. I told her to be standing at the door with my ring because I was coming to get it. And I went and got it and never went to see her again.

Even though my way of showing it was to drench her in cold water twice, I fell in love with Pat Leserra the very first time I saw her. It's a love affair that's lasted over 50 years. (**Courtesy of Donnie and Pat Allison**)

That was just the way I was back then. Stubborn, hard-headed and determined that nobody was going to tell me what to do, especially a girl.

Fortunately, Pat didn't let anybody push her around, either, not even a crazy Allison boy with racing on the brain. Fortunately, she didn't hold a grudge over me dunking her twice then being a jerk about it, either.

We started dating and six months later, I went to my mother to ask her for my birth certificate. She gave me a hard look and said, "You're getting married aren't you?" I told her I was, and she told me, in all seriousness, "It won't last two weeks."

That was about like Bobby telling me I'd never make a racecar driver. It meant I had to show her and for almost 50 years, I got the biggest kick out of teasing her by saying, "You said it wouldn't last two weeks."

Pat Leserra was a Godsend for me. When we got married, she was 16 years old and still in high school, but she was more of an adult than I was at 21. She had a better understanding at that time of what it took to be a grownup in a marriage.

It's a good thing, too, because we got married on Tuesday, and I left on Thursday morning to come back to Alabama for the racing season. I left Pat at her parents' house for all of about two weeks, then I called her and told her to get with Bobby's wife, Judy, get in the '61 Chevy I owned and get to Alabama.

They came with six-month-old Davey in the backseat, and so began her life as a true racing wife. When I look back on it now, it's a wonder it wasn't enough to send a teenage newlywed running home to Momma.

We shared an apartment with Bobby and Judy in Five Points, a suburb of Birmingham. We never had any money. There wasn't any such thing as credit cards back then, and I wouldn't have wanted her to use them if we'd had them. I was gone racing all the time, then I'd come in with a suitcase full of dirty clothes and expect her to wash them.

She always did, too. I could come in with every piece of clothing I owned stuffed in that suitcase, and two days later it would be packed with clean clothes ready for me to go again. It was a tough way for her to live.

That's one of many reasons I was so happy when my daughter, Pam, was born on March 2, 1962. It gave me the most beautiful little girl a man could possibly want, and it gave Pat something to fill the emptiness that had to be there when I was gone.

We had my son, Kenny, on Feb. 17, 1964, and not long after I packed Pat up and we moved to Chattanooga so I could drive for Bob White. At the end of 1964, we moved back to Alabama, and even though the house we had was a little bit crowded, we started talking about having a third child.

I told Pat I didn't want to stop at three, so if we had a third child, I'd want a fourth later on. Growing up in a big family like I did, I thought it was important for children to have buddies.

If we had stopped at three, chances were two of them would have been close and the other would have felt left out. So I felt it was important to have an even number.

The Lord must have heard those discussions because I was sitting at Fred Lorenzen's house in Illinois in late July when Pat called me and asked if I was sitting down. I asked her what the doctor said, and she said she was about to tell me, which is why she wanted me to sit down.

She said the doctor had heard at least two strong heartbeats, and this was before ultrasound, so that's all they had to go on. It gave me chills because I knew I was going to be the father of at least twins.

Two weeks before her due date, Pat had to go in the hospital. She had gotten really big, and I was surprised they didn't go ahead and induce, but the doctor said with multiple babies, he wanted them to stay as long as they wanted to stay.

On Aug. 16, 1967, Ronald and Donald arrived. Ronald came out first weighing 8 pounds and Donald came next weighing 7 pounds, 2 ounces. Pat weighed 49 pounds less after the delivery.

To show how obsessed I was with racing and what kind of constitution Pat has, I'll tell this story.

Right after the twins arrived, I went back to Illinois for some reason, and Pat went home from the hospital. Her mother was there to take care of her for a few days until I got back.

Three days later—less than a week after giving birth to twins—I got home, and Pat picked me up at the airport. I asked her what she was doing, and she informed me she and her mother were going shopping. Like I said, she's one strong woman.

From that point on, our lives, even with all the ups and downs of racing, the heartache and the pain, have been an absolute joy.

The Perfect Mate

As a husband, you know you've found the perfect mate when she'll go sit in a duck blind with you. Or climb on the back of a bass boat of her own free will. Or let you go hustle a game of pool down at the pool hall every now and then.

Pat did all that and never complained. In fact, she put up with all of it, and enjoyed most of it, except maybe the time the wild hog stole her lunch.

We were hunting dove one day, and Pat was sitting on a stool in her blind, gun across her lap and lunch in a brown paper bag at her feet. She hadn't been sitting there long when she heard a terrible racket behind her and turned to find herself face-to-face with a wild hog.

Like I've said before, Pat's a smart lady, so she vacated the premises quickly, leaving the stool and her lunch behind. Later, we went back for the stool and found the paper bag ripped to shreds and Pat's lunch gone. Old Mr. Hog had helped himself, and it looked like he enjoyed it, too.

Another time I took Pat with me down to south Alabama to go duck hunting in Mobile Bay. We were sitting in the surf in the Gulf of Mexico now, waders on and freezing cold, when a big old mallard flew over, and I got him. Unfortunately for Pat, he came down right beside her and splashed water all over her.

She didn't seem too unhappy about being soaking wet in the freezing cold, though. The only time I ever remember Pat being unhappy about anything along those lines was the time I stayed out all night playing pool.

When we first married in the early 1960s, we didn't have any money. We literally went from day to day, week to week, month to month on what I could scrape up driving a racecar. So to help out, I'd go to the pool hall and make grocery money with a pool stick, and I was pretty good at it.

There was a place close to where we lived in Bessemer called Pharo's, and I'd go in there a lot at lunchtime. The car salesmen and insurance salesmen would go there, and we'd play $1 9-ball or $1 bank or some game like that.

Just about every day I was home I'd go in there and maybe on Tuesday night if I wasn't racing. Just about every day I'd go home with some money, and often I'd make $30 to $40 in a week. It doesn't sound like much, but to us at that time, it was a lot—enough to help us survive.

One time, I stayed out all night playing pool, and it was already getting daylight when I came home.

I tried to go in the house, but the door was locked, which was unusual because we didn't lock our doors back then. This time Pat had locked it, though, and she told me to go back where I came from if I was going to stay out all night.

I told her I was going around back and if the door wasn't unlocked when I got back there, I was going to kick it in. It was unlocked when I got around there, but that made an impression on me.

It made me think, what was I doing out all night playing pool? I was making money, sure, but I had obligations at home. I had a wife and children I needed to be home with instead of out pool-sharking. Pat and I never discussed it again, but I never stayed out all night again, either.

There have been times in our lives when we've had "discussions" like that, but one thing I will swear to is that, in over 50 years of marriage, I don't ever remember us going to bed mad at

Pam and Kenny were growing like weeds, and the twins were well on their way in this Allison family portrait from 1967. (**Courtesy of the Morris Keene family**)

one another and getting up mad the next day, and that's pretty darn good.

You hear all the time about matches made in heaven. Well, that's the only place this one could have been made because if it hadn't been, with all the stuff we've gone through over the years, we wouldn't have made it.

A Close-Knit Family

I am close to all my children. Kenny has been bass fishing and hunting with me from the time he was six years old. He played football, and I even coached his team one year. He went just about everywhere with me.

I took Pam hunting with me several times, but she got into dance and that was her thing. Her mother did all the dance stuff with her, but I'd go to her recitals.

Ronald and Donald played sports growing up and were pretty good athletes, but they were into hunting and fishing big-time, too. We're all pretty crazy about the outdoors.

They're all good kids, they've worked hard and they've never given us any trouble. Not much anyway.

I do remember one time, though, when Ronald and Donald brought home their report cards from junior high school, and they were not good. Ronald had three failing grades and Donald had a failing grade and two Ds.

This was during hunting season, and the rule was they had to keep their grades up if they wanted to keep going to the woods.

They didn't want to show me those report cards, and I understood why when I saw them. I took one look at those grades and went through the roof.

I told them to go in the den where we had a pool table and I'd be in to "talk" about it in a few minutes. I knew I couldn't spank them right at that moment because I might really have lost my temper and hurt somebody.

A few minutes later I went in and pulled off my leather belt. I said, "OK, which one first?" Ronald stepped up and said he'd go first, and I hit him two pretty good shots, probably harder than I should have hit a ninth grader, to be honest.

I asked him if that was enough and he said, "Yes sir, I think that's enough."

Then Donald put his hands on the pool table, and I gave him two shots equally as hard. Then I said, "Is that enough?" And he said, "No sir Dad, I think I need one more." So I jacked him off the floor about six inches that third time.

Donald wasn't being smart, though. I think he knew better than to be smart with me at that time because I was really hot. No, I think he realized he needed that third shot because he had not lived up to his capabilities, and that had disappointed me and his mother.

It was the last time I ever had to discipline any of my kids, and I have to say, they've all made their old pop pretty proud.

Today, we all live within a few miles of one another in Salisbury, N.C., and Pat and I help the boys in their business, the Allison Legacy Race Series.

Kenny and Ronald and Donald build the Legacy cars as we call them, scaled-down versions of real racecars that younger drivers—and some not so young—buy to get their start in racing. Pat works with the boys doing the bookkeeping and scheduling for 20 to 25 events across the country that make up the Series season each year.

Pam is married to former successful short-track and Winston Cup driver Hut Stricklin, who owns his own thriving junkyard business in Salisbury.

Between them, they have given us nine grandchildren (Stephanie, Samantha, Justin, Sarah, Taylor, Tabitha, Hannah, Natalie and Josh) and four great grandchildren (Cade, Kylie, Wyatt and Silas). Of course, as grandparents do, Pat and I love spoiling each and every one of them.

The entire Donnie Allison clan turned out for the ceremony marking my induction into the International Motorsports Hall of Fame in Talladega, Alabama, in April of 2009. (**Courtesy of Jimmy Creed**)

AS I RECALL...

We're so close, and it's so nice being as close together as we are in Salisbury that it probably means our chances of ever moving back to Alabama again are pretty slim.

I love Alabama and have since I first set foot there over 50 years ago. If I had my way, I'd build a big house on the farm in Faunsdale and Pat and I would move back there permanently. I'd ride my tractor, cut my hay, tend my cows and be very content.

But Momma loves being close to her children and grand-children in Salisbury, and I know her well enough to know she's not moving anywhere too far away from family, even back to the farm in Alabama. And I'm not going anywhere without her, so I guess we'll stay put right where we are and grow old together.

That Banjo Sounded Pretty Good

I got my start in Grand National racing in 1966 when I drove two races for a man named Robert Harper out of Jackson, Miss. But it was 1967 when that part of my career really took off.

I felt so good going into my first full season in Grand National racing that I predicted beforehand I would be rookie of the year, then backed it up. I was never afraid to take on a challenge, and I did there because I had all the confidence in the world I could do it.

Harper was a good man, but he couldn't decide if he wanted to do it full-time or just play at it, so after five races in 1967 I moved on to a ride with J.D. Bracken and then Jon Thorne. Between those three, I drove 20 races that season, and while I didn't win one, I sure made an impression.

Back then the media voted a rookie of the race after every race, and I won it in every time I ran. At the end of the year, they voted me the Regal Ride by Monroe Shocks Rookie of the Year, too.

That impressed enough of the right people to earn me a shot at a big-time ride in 1968.

I went to Daytona on a one-race deal with the Holman-Moody team, which was the Roush Racing or Hendrick Motorsports of my era. I qualified their Ford seventh and ran good, but I cut a tire down, put it in the wall and tore the car up bad.

I was just sick about it because I figured my first real chance had literally met a stone wall, so I was expecting the worst when John Cowley sent word for me to come to the Holman-Moody office there in Daytona. Cowley was one of the big bosses in Ford's racing program, and I didn't think he liked me because I knew he didn't like Bobby.

But Cowley surprised me. Instead of firing me, he offered me a one-race deal with "Banjo" Matthews' team for the race coming up at Rockingham. There was a catch, though.

NASCAR had just allowed teams to downsize to 390-cubic inch engines instead of 427s to give the teams a break on the weight. All the big Ford factory teams had the 390s, but Banjo didn't.

Cowley told me I could run with "Banjo," but I'd be running a 427 instead of a 390, which meant I was at a disadvantage from the start, but I didn't care. When he asked me if I wanted the job, I said "Yes sir!" fast before he could change his mind and kick me out.

Cowley said "Banjo" was going to test in Atlanta and wanted me to go with him. So I drove "Banjo" from Daytona to Atlanta, and that was truly the start of my big-time racing career.

"Banjo" Matthews was contrary, ornery, very set in his ways and one of the best racing men I ever came in contact with.

Over the years, he prepared cars for Fireball Roberts, Junior Johnson, Cale Yarborough and A.J. Foyt among others. Yes, I followed A.J. in that ride just like I later followed him in the Wood Brothers' ride and with "Hoss."

Because of their personalities, A.J. and "Banjo" didn't get along, and there were probably plenty of people in the garage who didn't think "Banjo" and I could make it, either. But after that first Daytona-to-Atlanta trip, the old "Banjo" always sounded pretty good to me.

I had heard of "Banjo" before that day, but I didn't know him. He had run some down in Opa Locka and Hialeah in Florida where Bobby and I got started, but I'd never had any dealings with him.

We hit it off because we turned out to be very much alike. There was no B.S. about "Banjo," and I had always been straight forward, too. But being around "Banjo" made me even more so, a lot of times because I had to be in my own defense.

"Banjo" helped mold me into the person I am today because he was no-nonsense, blunt and right about almost everything he taught me about racecars and life.

On that trip to Atlanta, I told "Banjo" I needed help because I didn't really know how to drive a Grand National car yet, even though I had been rookie of the year. I told him I still needed to learn how to do it and wanted him to help me.

I've never forgotten what he said to this day. He looked me straight in the eye and said, "Donnie, it's a racecar. Just drive it."

All these years later, I can truly say it's as simple as that.

Driving a racecar is just like flying a plane. It's not going to do anything you don't make it do. So you've got to make it do what you want it to do.

If you want it to turn, you've got to turn it. If you want it to speed up, you give it the gas. If you want it to slow down, you hit the brake. And you can't get all tensed up while you're doing it.

That was what "Banjo" was trying to tell me. You can't be nervous, uncertain or scared when you're in a racecar. If you are, big trouble is just around the next turn.

That's stuck with me ever since because it's so true. I get a chance to talk to a lot of youngsters trying to get started in racing through the Legacy Racing Series and just like "Banjo" did, I tell those who come to me for help, "It's a racecar. Just drive it."

A pit stop during a race at Rockingham with "Banjo." Yes, that is the car owner changing the front tires during the race—something you would never see these days. (**Courtesy of Donnie and Pat Allison**)

My partnership with "Banjo" got off to a stormy start because of stormy weather. We got rained out in Rockingham, so they asked me if I wanted to go to Bristol the next week on the same deal, and I said yes again.

We went to Bristol with the 427 engine, and I qualified sixth, but finished 30th because something in the rear end went out. Things really started looking good, though, when I finished third at Atlanta and Martinsville in the next two races I drove for him, then came in 29th at Darlington.

We sat on the pole together for the first time at Charlotte in May of 1968 then went to Rockingham and got our first win together, one that was very emotional for both of us for very different reasons.

It was blistering hot that day at North Carolina Speedway as I beat Bobby and the rest by two laps driving the No. 27 car

"Banjo" and I got our first win together on a blistering hot day at Rockingham in 1968. (**Courtesy of Donnie and Pat Allison**)

with a 427 engine. But the image that will be burned into my mind forever is of "Banjo" crying like a baby in victory lane afterwards.

I was crying tears of joy, for sure, but "Banjo" was crying tears of relief. He told the reporters afterwards he hadn't really expected to get another chance with Ford for the 1968 season and was pretty sure if he'd failed to win any races, he would have gotten the axe at the end of the year.

He later told me he'd been so beaten down by the deal with Foyt in 1967 that he'd never expected to win another NASCAR race and was resigned to getting out. Instead, our relationship just blossomed after that, and we won four more together over the next two years.

"Banjo" was one of the two biggest influences on my racing career, and the other was Eddie Allison.

Nobody is smarter about racing than Eddie, and he was very instrumental in helping me learn to drive a racecar. The way he talked to me, the way he treated me, the way he helped me was the best.

He was never bossy. He never ordered me to "Do this" or "Do that." It was always, "Why don't you try this?" or "Let's try that."

Eddie's only shortcoming in racing was that mentally, he couldn't lay it down. He was so consumed with it that he couldn't get away from it, and it made him physically sick.

He did the same thing in his construction company he later worked for. He was so consumed with doing a good job and running the business, that it ruled him 24 hours a day, seven days a week. But he was good at that as he was at racing, too.

"Banjo" was exactly the same way. He never ordered me to do anything. Never threw his weight around because he was the owner. Never pulled rank.

I hear that way too much today with car owners and crew chiefs. They tell drivers to go do this or that and never stop to listen to what the driver has to say or ask him what he thinks. You can't do that if you want to have a good racecar driver, and thank goodness, those men never did it to me.

"Banjo" was a mentor. We would talk and in his certain way, he would tell me things I would later find to be unbelievably accurate. But he knew because he drove.

I had all the respect in the world for the man. I realized pretty quick that, if I wanted to sit and listen, I could learn as much as I wanted to from this genius, and I soaked in as much as I could.

Being able to watch a racecar and tell what it's doing and what it needs to be better is a gift not a lot of people have. My brother had it. "Banjo" had it. And I was truly blessed to have them.

A Good Dance Partner

When I first got into Grand National racing, as it was called back in 1966, other drivers quickly told me to watch out for Buddy Baker.

Buddy had a reputation for holding it wide open whether he was going forward, backwards or sideways. They said there were only two ways he knew how to drive, wide open or shut off.

It made him a rough driver. He didn't have a lot of finesse. He just manhandled a car where he wanted it to go, regardless of where that was or who might already happen to be there.

I didn't really understand what they were telling me at first, then I started practicing with him, and I saw very quickly what they meant.

In those days, we practiced a lot. We'd usually come in a day or two early and run one full day of practice. We'd have a two- or three-hour practice in the morning and another two- or three-hour practice in the afternoon. We ran together a lot. We drafted a lot, which is not something they do today.

Nobody wanted to run with Buddy, though, because you couldn't predict what he was going to do. He might run the top one time, the bottom the next and keep you guessing about it the whole time. He was just difficult to work with.

Buddy asked me one time, "Why doesn't anybody ever want to run with me in practice?" I looked him straight in the eye and said, "Because you're crazy." He laughed and made a big joke out of it, but I was telling him straight out what we all believed.

When it came to working with people on the track, Richard Petty was the best. Richard would sit back there on your bumper and run with you as long as you wanted him to because he was accomplishing something himself.

He realized it wasn't about leading the race at the 10th lap or the 50th lap or the 100th lap. The idea was to lead the race when the checkered flag flew.

Darrell Waltrip was the same way. He would run with you for 10, 15, 20 laps, and the only time you'd swap positions was for you to cool your engine or him to cool his. He didn't put you in a bad position just to lead the race for no apparent reason.

David Pearson was in that group, too. Pearson would sit in third or fourth or fifth and be content to run there all day because he knew he was driving for the Wood Brothers and they had some of the best pit crews in the business.

He knew when he stopped he was going to be back out second or third every time, and then he could go win the race. So he would be patient and work with guys early in the race because he wanted to be there at the end. That strategy helped David to 105 wins, second-most in NASCAR history.

There were guys that were difficult to work with, too, like Cale Yarborough and even Bobby. Honestly, I never could draft with my own brother like I would have liked to. We worked together sometimes, but not always.

The problem was that Bobby was very, very self-centered on the track. What I mean is that even if Bobby tried to work with somebody, he never wanted to be behind. Ever. That made it very difficult to give and take in those situations because Bobby always wanted to take and never give.

Cale was just like Bobby. You could run with him sometimes, but he never wanted to be behind, either. So if you wanted to win, you had to be careful if you picked one of those guys for a dance partner.

One of the best dancers out there in my day was … well … me. Honestly, I don't think any of those guys drafted any better than me.

If you pay attention when you're driving a racecar, if you watch what the other drivers are doing, it's an advantage. I watched them closely, I understood what I was doing, and it helped me run up front a lot.

Some of those guys would argue with me saying I was the best drafter of my day, I'm sure. But there's one thing I think all of them would admit, even now. Whenever I was in a car capable of running up front, everybody wanted to run with me to get there. And I mean everybody.

Talladega and the PDA

In the fall of 1969, "Big" Bill France was set to open his huge, new racetrack in Talladega, Ala. The problem was we didn't have a tire we could race on, or so we thought.

This led many drivers to boycott the first race there in September of 1969 and to a defining moment for NASCAR as an organization.

I was the first racecar driver ever to drive around Talladega. Herb Nab, who worked as a crew chief for "Junior" Johnson, had jumped in a racecar earlier and driven around it slowly, even though he wasn't supposed to. But I was the first driver ever to get a racecar up to speed there, and what I saw shocked me.

I went to a Ford test there in August of 1969, and the first time I went out, I was in awe of how much room there was. Even though Talladega is only a tenth of a mile longer than Daytona, it's 12 feet wider. Twelve feet may not sound like much, but it makes an unbelievable difference in what it's like driving around the two tracks.

I was very disappointed, though, because it was so rough. I went to run a competitive speed for the first time, and my car bottomed out. I hit a bump going into turn one, and it almost lifted me off the ground. I came back down and bottomed out again. The racetrack's surface in the middle of the corners was terrible.

The tires we had at the time were Goodyears, and they weren't good enough. The tire companies—Goodyear and Firestone— had to go back and do a little work before they came back for the race.

The race weekend came, and I was going to run Firestones, one of the few drivers actually on their tires. I'll never forget how, in two laps of qualifying, I tore up all four tires. Several of the Goodyear guys blew tires, too. Not everybody, but a lot of them.

What happened was the racetrack got finished in typical hurry-up fashion. It really needed another month or two to season, but Bill France had a deadline to meet for his creditors, and it had to open.

Well, the tire companies didn't have sufficient time to make and test a tire for Talladega, and they didn't want to have a problem. They were the last people who wanted to have a problem because it was bad advertising.

Had they had the time and the proper kind of testing to sort out the problems, Goodyear and Firestone would have built tires we could have raced with. They both had the capabilities; they just didn't have the knowledge or the time to do it.

When I left Talladega on Thursday, September. 11, the Firestone people had taken their tires up, and I heard Goodyear was taking theirs up, too. So I was under the impression they were going to do something to try to stop the race.

The next day in the garage sign-in area, Ed Alexander, the head Goodyear guy, and a couple of others were there. He spoke and shook hands, then got down to business.

He said he thought their tire would run. I said I wasn't so sure because an awful lot of them had torn up. Then he asked me to run 20 laps to test it for him.

I asked him if we were on test insurance, which was very good. He said no, we were at a race. At that time, NASCAR didn't have very good insurance. We ran anyway, but we really weren't too bright for it.

I told Ed no and to get one of his insured drivers to do it. So he got LeeRoy Yarbrough to go out in one of "Junior" Johnson's cars.

The first time out, he ran three laps and came in and told them he felt something wrong. They checked it out and nothing was wrong, so they sent him back out. He ran eight laps the second time and came in and told them something was tearing up.

LeeRoy never did run a full 20-lap test, and from that time on, things got considerably worse. Guys decided they weren't going to drive, and it really became a fiasco.

What we didn't know because nobody told us was Bill France had a lot of pressure on him to satisfy a lot of people—the money people, everybody that was behind building the racetrack.

As a racecar driver, all I was concerned about was not having a tire that would stay on my car. But it was not that simple.

Bill talked to us in the garage before it really got out of kilter and told everybody standing there—but he was looking straight at me when he said it—that nobody was telling us to hold it wide open.

I asked him if Richard Petty passed me in a car he owned and my foot wasn't to the floor, wouldn't he want me to hold it wide open? He said we would have to decide that for ourselves, which was not a very good answer to me. After that, it really blew up.

There's a picture of me standing looking up at Bill France, a 6-foot-6 monster of a man, questioning him about his duties as the president of NASCAR. I told him I thought the president of an organization should look out for the welfare of the organization or something to that effect.

He looked right at me and said, "You're individual contractors. NASCAR doesn't have anything to do with it." So I asked him why I had to have a NASCAR license. I don't remember his exact response, but it wasn't too nice.

I was pretty well spent by this time. Here I was a young man driving for the Ford factory team and desperately not wanting to lose my job. But I'd made a pact with myself that I could not run because I knew I was going to kill myself or somebody else.

Fortunately, the Ford folks didn't force the issue with us. They called each of their drivers and car owners to a meeting at the Holman-Moody truck.

One at a time, we went up on top of the truck to talk to the Ford boss, Charlie Gray. The only question he asked was "Are you going to run this race?" I looked him straight in the eye and told him I hoped he didn't have to fire me, but that I couldn't run the race and then got down.

"Banjo" went up next and Charlie asked if he was going to run the race. His answer was "Only if Donnie Allison drives my car," and then he got down.

It made me feel good because he was letting me know that they weren't going to tell him to put somebody else in the car, and they didn't. There were never any repercussions from Ford over the incident whatsoever.

After that, things quickly got out of hand. People started talking out of turn and arguing. Petty said something to Bill and even brought up the Professional Drivers Association (PDA).

Earlier that year, a group—including me, Richard, Bobby, and David Pearson—had gone to New York to meet with Larry Fleisher, an attorney and general counsel to the National Basketball Association players union. When we talked with him in his office, the first question he asked us was about money. We told him we didn't want to get into the money. He said, "You will."

We didn't want to form the PDA for money. What we had in mind was simply to get enough collective bargaining power to get some improvements made in facilities for ourselves and our families at the tracks.

At the time Talladega was built, there was very little in the way of facilities in any of the garage areas. Talladega had a tiny bathroom in the garage that had a two-stall shower with no divider. It had a great big urinal down one wall and a couple of commodes.

This was where we were supposed to clean up and change— no locker rooms, no privacy, no nothing, and it was even worse for our families.

When our wives and kids went to the races, they had to go in the infield where it was really rough. They had to use port-a-johns or whatever they could find. They had to deal with the drunks and everything else that went on there. This is what we wanted to change, and that's what the PDA was formed for.

Fleischer told us what we were trying to do was good, but it wouldn't be long before money became the main issue, and that really sunk in with me. I decided that was true because every time I heard about a union, that's what came up. It wasn't the welfare of the people; it was the money.

We formed the PDA and tried to make it work, but with God as my witness, it did not have anything to do with us not running at Talladega that day.

There's a big misconception that we went on "strike" that weekend, but it is simply not true. The PDA did not voice an opinion. It didn't threaten anybody. I know. I was one of the PDA higher-ups, and I know none of us made any kind of threat about not running that day or any day.

Truthfully, the reason everybody went home was because Bill told Richard to get out.

We were talking in the garage area, and Bill looked around and saw Richard's car sitting on the ramp, about halfway into his drive-on truck. He turned to Richard and said, "If your car is loaded up already, go ahead and get out of here."

When he said that to Richard Petty, the cars hit the trailers. I've never seen anything like it in my life. It made a lot of people mad because they felt like he was throwing Richard out, which he really wasn't.

Bill was just frustrated at the time and understandably so. Here was a lifelong dream to build this racetrack and he had all this money involved, and the main drivers decided they weren't going to run. I can understand why he got mad.

We all loaded up and went home, and Bill ran his race the next day, filling out the field with sportsman cars, back markers and journeymen drivers like Richard Brickhouse, who won the race and was never heard from in racing circles again.

I didn't understand it at the time, but I'll admit now, the best thing that ever happened in the entire history of NASCAR was for Bill to run that race.

It showed that he ran the show and that you were going to do like he said if you were a member of NASCAR or were going to run a NASCAR race. He didn't buckle. He didn't back down, even though I felt like maybe he should have.

He told us standing there in the garage that day he was going to run a race, and we were skeptical of that. But he showed us. Right, wrong or indifferent, he ran that race.

It's why NASCAR is the ultra-successful organization it is today. There's one chain of command, one man at the top who ultimately makes all the decisions and there's not a driver's union. It's about as simple today as it was in 1969. Either do it France's way, or get the hell out.

A Taxi Driver at Indy

I always used to ask A.J. Foyt when he was going to let me drive one of his Indy cars. In early 1970 at a Goodyear tire test in Phoenix, he finally took me up on it, and it cost him $500.

A.J. bet the Goodyear guy running the test I wouldn't break 30 seconds on any of my laps, but he didn't tell me about it. I got up to speed and my first timed lap was 30.1 seconds, and my next one was a low 29.

The next lap, I came around and there was A.J. standing out on the racetrack on the front straightaway. He had climbed over the guardrail and was waving his arms for me to come in.

When I stopped, he came running up and yelled, "What are you doing you SOB? You just cost me $500." Of course, all the Goodyear guys were jumping up and down, hollering and laughing because they liked seeing A.J. lose money.

If he'd told me, it might have been a different story, but since I didn't know anything about it. I just ran the racecar like I was supposed to.

He didn't think I'd run that well, but it wasn't meant as any disrespect to me. It was just A.J.'s way.

That's me in a 1968 Dan Gurney Eagle owned by A.J. Foyt that I drove to a fourth-place finish in the 1970 Indianapolis 500. **(Courtesy of Donnie and Pat Allison)**

I fooled him several times in a racecar like that, though. Two of those times were when I ran the Indianapolis 500 for him in 1970 and 1971.

Very few people remember I was the first stockcar driver to successfully double dip in NASCAR and Indy car racing. Long before John Andretti, Robby Gordon and Tony Stewart, I did the

Memorial Day weekend double at Indy and Charlotte and did it better than anybody before or since.

On May 23, 1970, I finished fourth my first time out at the Brickyard in a 1968 Dan Gurney Eagle. Then I jumped in a jet, flew to Charlotte and won the World 600 the next day when it was still run on Memorial Day itself. Then I flew back to Indy to attend the winner's banquet that night where I accepted rookie of the year honors.

The next year, I went back to Indy and finished sixth in another one of A.J.'s cars.

It was fun, but those were two hellacious years with A.J. at Indy and races we ran at Ontario, Pocono and Milwaukee.

Let's just say A.J. didn't give me a lot of help. By that I mean he didn't talk to me very much about how to drive Indy or those other tracks. Instead, he belittled me, made me out to be the little guy to everybody, gave me hell everyday about being a taxi driver.

It was kind of a comical thing, and I didn't let it get to me. If I had, it would have driven me nuts.

Actually, my Indy career almost ended before it ever began.

I went to the orientation for my rookie test in 1970, and they told me I had to run through a series of lines painted on the track apron. They told me during the first part of the test I had to run those lines every time.

In the garage, A.J. told me he didn't care what they said, I shouldn't run those lines under any circumstances. Well, I tried to do what I was supposed to do for the United States Auto Club officials, and I ran the lines and spun out.

When I came in, A.J. was really mad at me, so mad he went and got in a big war with the USAC officials about making me run those lines. I know because I heard him from all the way down in the garage where I was.

DONNIE ALLISON

Pat, Kenny and I pose with the plaque I received for earning rookie of the year honors in the 1970 Indianapolis 500. **(Courtesy of Donnie and Pat Allison)**

Believe me, I was concerned. I didn't know if I was ever going to run another lap at Indy in my life. I didn't know what they were going to do to me.

A.J. came back and said the USAC officials wanted to see me. I went down there, and they told me not to run the lines the next time. So I went out, finished my rookie test and came back to the pits.

To qualify, I had to run one phase of the test at a certain mile an hour, another at a certain mile an hour and a third phase at another. I finished the first stage and had two left. But they told me to go on out and run. Spin and all, they'd seen enough in just one phase for me to pass my rookie test.

A week later, I made a rookie mistake and crashed the car.

I learned very early that when you came down the back straightaway at that time you had to look at the wind sock to see which way the wind was blowing and how hard. I didn't do that.

At that time, there weren't any seats down the back straightaway coming off turn two. It was open, and I could see the golf course. There was a gate there, and when I got to that gate opening, a gust of wind caught me, and I did a 360-degree spin.

I hit the wall nose first going straight ahead and hit it hard. I bumped my knee and bit my tongue, which was bleeding pretty good.

A.J. was in the garage telling everybody he told me I was going to wreck. He hadn't said anything like that to me. We had put a new set of wings on the nose of my car, and if he had told me to be careful because it was going to make the car's nose stick real tight, it would have been a good reminder. But he hadn't.

I came out of the infield care center, and he asked if I was all right. I told him yeah, but I had bitten my tongue. "Bit your tongue?" he said. "How'd you bite your tongue?" I looked him straight in the eye and said, "Because just before I hit the wall, I said, 'Oh shit!' and I bit it."

Race day was easy compared to the month leading up to it.

I started 23rd, and as the 200 laps wound down, I traded places with Bobby Unser and Mario Andretti for fourth, fifth and sixth positions. We passed each other about five times in those last few laps, and I finally got fourth, and they couldn't get it back.

I actually thought I ran third, but they put Dan Gurney in front of me. Still, I was tickled pink. Of course, A.J. didn't say "Good job" or anything like that. He didn't win, which made him so mad he wouldn't talk.

A.J. Watson, Unser's crew chief, did talk to me in the garage afterwards, and I'll never forget what he said.

I was sitting on a bench changing my clothes, and he walked over to congratulate me and shake my hand. He said, "You're the only stockcar driver I've ever seen who could drive these cars." He didn't have to do that, but he did. That's why what he said was so important to me.

★ ★ ★

In 1971, I had a Coyote that A.J. had run the previous year, and I was really struggling. It just wouldn't run. But all he could say was "I ran 172 miles per hour in that car last year."

We got in a pretty big argument, and he gave me that "I ran so-and-so" again, and I told him to do it. Show me. He started saying I had done this and that to the car. So I undid it.

I put the pedals back. (In an Indy car you adjust the position of the pedals, not the position of the seat.) I put the steering wheel back. I had cut a quarter-inch off the windshield, and I couldn't do anything about that, but I told A.J. I'd give him a mile an hour for that.

He got in the car, and he was going to show me. He ran his laps and came in. He raised his face shield, looked up at me and said, "See."

Well, a guy that worked for him named Jack Starnes took the speed board, handed it to him and said, "You better see." He didn't come within a mile-and-a-half of being as fast as I was in that car.

A.J. got out all huffy like he always did and went to a new car he had sitting there with a Goodyear blanket on the nose. He told me to fold up the blanket and go for a ride.

So I got in a new car, one I had never run a lap in in my life, and the first lap I ran 173.5 miles per hour, which was faster than he had run it.

Long story short, he gave me that car, and I was just about to take it out for a second time when the caution light came on. One of the kamikaze guys that come out on the last day of Indy qualifying had hit the wall, and I had to head in instead.

As I came in, some of the guys came running down pit road motioning for me to get in the qualifying line. A.J. had decided he was going to withdraw the car I had posted my first qualifying time in—one that was sure to get bumped— and have me re-qualify in this car that I had run a total of one lap in.

This caused a big cuss fight between A.J. and Andy Granatelli, who were always trying to outdo each other at Indy. As usual, A.J. got his way, and he finally came over and knelt down beside me in the car.

He told me to run as hard as I could off pit road because they were going to time me off turn four the first time. If that time wasn't good enough, A.J. planned to throw the yellow flag, waving off the qualifying attempt.

I told A.J. he could stick that yellow flag were the sun didn't shine because if that car didn't feel right, I wasn't driving it.

On my first lap, they couldn't figure out how fast I was going. They said somebody even hollered out I was going over 175 mph.

They had told me that if I saw any kind of blue magnetic number go up on the board to raise my hand and take my time.

They didn't have to tell me that. I had a tachometer. I knew what the RPMs were. So I raised my hand and took my time. On my four laps, I ran 174-something, 174-something, a high 173 and a 172.8, good enough for the 20th starting spot.

When I stopped in the pits, A.J. didn't say anything nice. All he could say was, "How in the hell come you slowed down so much on your last lap?"

After the 1971 race, I told A.J. if he wanted to win a fourth Indy 500—which I knew he wanted desperately—he needed to concentrate on running the big race and let me run the full schedule.

He said, "You want to do that?" and, at the time, I did. If A.J. had wanted me to drive for him then, I would have walked away from stock-car racing, and who knows what might have happened.

I am very proud of my Indy accomplishments because I knew when I went there I could drive there. Nobody else thought stock-car drivers could do it, but I showed them, which is why what A.J. Watson said meant so much to me.

How good was I at Indy? Good enough that, after I completed all 200 laps in 1970, another rookie didn't accomplish the feat again for 23 years, until Nigel Mansell finished third in 1993.

There were stock-car drivers who tried Indy before me and many since. But if you measure it by the circumstances I went under, the conditions I raced in and my performance in the end, there's no doubt in my mind I was the best of them all.

The Greatest Driver Ever

A.J. was brash. He was arrogant. He could be a bear to deal with. He was also the best racecar driver I ever saw.

I've heard people compare lots of drivers over the years, and one name that always comes up when they talk about who's the best is Mario Andretti. Andretti did this. Andretti did that.

Andretti was a marvelous racecar driver. But he was nowhere near as good as A.J. Foyt.

Everything A.J. ever ran in, any type of car, any circuit, he won in. He won in stock cars. USAC cars. He went to Lemans and won it. If he had gone to Formula I like Mario did, he would have won in that, too.

One time I saw Foyt take his old midget out of mothballs and beat the best midget racers in the world at an indoor race in the Astrodome.

True to his reputation, A.J. Foyt was often difficult to deal with. But he also helped me on several occasions throughout my career and, in my opinion, was the best racecar driver I ever raced against. (**Courtesy of Donnie and Pat Allison**)

In 1971, they put a dirt track down and brought in Bill Vuckovich, Tony Bettenhausen, and a bunch of others with their sloped-body midgets, which were the latest design at the time.

When the race started, the only way anybody could get the lead was to go out of bounds below the tires set up in the corners. That was the only way they could pass A.J.

So USAC stopped the race and said that anybody going under a tire in the corner would be disqualified. Foyt said to them then, "If y'all stay in the racetrack, y'all ain't going to keep up with me."

He was right. He blew them into the woods. It wasn't even a contest. He looked like a gorilla riding a tricycle. He was so big, and he was wadded up and stuffed inside this little midget. Let's just say there was plenty of him hanging out. But he blew away those great midget racers in an antique.

Another guy that deserves mention in any best-driver debate is Bobby Allison, and not because he's my brother. Bobby deserves mention because he was another driver that won in everything he got in and won consistently.

He won on big tracks. He won on small tracks. He won on road courses. He won on short tracks in modifieds and sports-man cars. He was the fastest-qualifying rookie ever at Indy in 1973. Everything he ran in, he was good in.

And he did it in spite of himself.

Bobby was a head's-up driver, very much in the same category as David Pearson, who won 105 NASCAR races in his career. The difference was that Bobby's temper would get away from him, and David's wouldn't.

Bobby was his own worst enemy with his temper and his ego. Always. For example, if he had kept driving for Junior Johnson, there's no telling how many more races—and championships— he would have won.

Cale won three straight for Junior in 1976, 1977 and 1978. Darrell Waltrip won three more for Junior in 1981, 1982 and 1985. If Bobby had stayed with Junior instead of getting mad and telling Junior he'd had enough after the '73 season, he could have won all of them and maybe more.

I tried to talk to Bobby about it several times back then, but all I got was a deaf ear. Besides, given the way we were at that time, it didn't make any difference what my opinion was; his was always different.

Knowing Bobby like I do, though, I'm sure he'll agree completely with me saying now that he's one of the best ever.

Bobby was good. He won 84 races, which ties him for fourth-most in NASCAR history with Waltrip. But he wasn't quite the best. That was A.J.

Some other guys on the list would be Pearson, Parnelli Jones, Roddy Perry, a modified driver I knew in South Florida, and Andretti. But I still remember Andretti coming to Daytona and winning the 500 in 1967 and then coming back in 1968 and almost not making the race in the same factory car.

I'm not saying Mario wasn't a good driver because of it. He was tremendously good. But the No. 1 racecar driver of all time? Not Mario—or anybody—over Foyt.

DiGard Disaster

Everybody remembers the No. 88 Gatorade Chevy Darrell Waltrip drove to 27 wins from 1976 to 1980 for DiGard Racing. What they probably don't remember is that I supplied the equipment, the know-how and most of the elbow grease to start DiGard Racing, only to get cheated out of everything.

Mike DiProspero and Bill Gardner were brothers-in-law from Connecticut who wanted to own a race team and had the money to do it. Bill ran a company that dealt in precious metals, so they did OK financially.

At first, they wanted to have a team with Bobby and me. We even had a meeting with Bobby in California, but he made it clear he did not want a teammate. Not even his brother.

I told Bobby he could pick his side of the shop and do his thing, and I'd do mine, but he wasn't interested.

So me, DiProspero, and Gardner formed our team in my little two-car garage in Hueytown. (A lot of folks thought DiGard was a company that made batteries or spark plugs or something for cars, but it just stood for Di-Prospero and Gard-ner.)

To get us started, I put up all my equipment and my shop in Hueytown in exchange for stock in the company, and they made me the president. It was one of the biggest mistakes of my life.

I had a 600 drive-on truck I had bought from "Banjo," two racecars, engines, pit equipment, everything. And I didn't owe on any of it. It was all paid for because I didn't believe in credit. Whenever I could make enough money to buy something, I bought it. If I couldn't, I didn't.

I put all this up because I believed the line of bull Bill Gardner fed me. I had a funnel up my butt, and Bill steadily blew smoke up it all the time. It was stupid I know, but I was young, and it sounded too good to be true. Maybe that should have been a clue.

I did it, though, because I had all the confidence in the world in myself, and I still do. What I didn't realize was I couldn't do it without the right kind of help, and I certainly didn't have that.

We worked out of the Hueytown shop in 1973 and part of 1974 and had our ups and downs.

We should have won the 1974 Daytona 500, but I had that blowout and finished sixth behind Richard. There was a second place at Talladega and a third at Atlanta in 1973 and a second, a third and two fourths in 1974. But we had a lot of really bad finishes, too, mostly because my crew chief, Mario Rossi, worried more about setting the car up for two laps of qualifying than 500 miles of racing.

If Mario had spent as much time trying to get the car to complete 500 miles as he did trying to sit on the pole, there's no telling how many races we would have won. Mario didn't care about winning races, though. He just wanted people to consider him a hotshot engine builder like "Smokey" Yunick and, for some reason, he thought winning poles impressed people more than winning races.

To be honest, Mario was the biggest backstabber I ever met. I didn't realize it at the time, and I would have probably fought you if you had told me that early on. But I learned that he fooled a lot of people for a long time in that regard.

Early in 1974, Bill came to me and told me he wanted to move the operation to Daytona. He said he wanted to be closer to the Frances and NASCAR. I knew what he was thinking, and I told him it wasn't going to make any difference if he moved closer; he wasn't going to get preferential treatment from NASCAR.

But he was determined to go, so we got a building on Fentress Boulevard, about a mile from Daytona International Speedway. It was a big building, and I went down and went to work on it.

My father came up and helped me install the lifts because that had been his business all his life. We put in the plumbing and all the air lines. I built the paint booths, and the engine room and it was nice. It was considered state-of-the-art at the time and got written up and photographed for a lot of the racing magazines.

We moved into the new shop me and "Pop" had worked so hard to build in July of 1974. One year later, Bill Gardner called me to his yacht at the Daytona yacht basin and fired me.

I sat on the pole for the 1975 Daytona 500, but there's a funny story behind that because it wasn't one of Mario's motors that put me there.

We'd had an open house at the new shop and an old-time drag racer named Bill "Grumpy" Jenkins came. I had a real nice, plush office there and somehow "Grumpy" ended up in my chair with his feet propped on my desk.

"Grumpy" was pretty well lit, and I didn't know him that well, so I took everything he said with a grain of salt. For some reason, he looked at me and said, "I like you. I'm going to build you an engine."

Just before the conversation ended, "Grumpy" said he would have somebody call me to get the engine specs. I thought, "Yeah, yeah, I'll believe it when I see it." Sure enough, the next day a man named Joe Tryson called me to get the specs so they could build my motor.

All Mario could say was he was going to build a motor we could use to show those drag-racing guys how it was done. So he built the best motor he could come up with, and we ran it at the track and ran good. We were really pleased.

But when we got back to the shop that day, Joe Tryson was there in a white van with the motor he and "Grumpy" built. So he took our motor out and put his in.

The next day we went back to the racetrack, and Mario just couldn't wait to show those drag-racing guys. I told him, "Mario, don't say nothing. Let's see what happens."

I was ready to go out the first time with his motor, so I asked Joe if he wanted me to give him a plug check.

To do that, I would run the car wide open, then mash the clutch, let off the gas and shut the engine off. That gave the engine man a true reading on the spark plugs at the wide-open throttle speed and told him if the fuel mixture and timing were right.

Joe said he was fine. I told him Daytona could blow up a motor in half a lap, but he told me again he was fine.

Well I ran, and I couldn't believe what I saw on the tachometer. My very first time on the track with that engine, I ran a mile-and-a-half faster than anything we'd posted the day before.

I came in and Joe asked a couple of questions. He asked me to go run again and this time to give him a plug check. I was a little more confident that the engine was not going to blow, so I opened it and ran another half a mile an hour faster.

By then, Mario was fit to be tied, but Joe Tryson wasn't through. I came in, and Tryson went under the hood, took one spark plug out and looked at it. He went in this little tackle box he had with him, took out his timing light and changed the timing on the motor. Then he told me, "That's all I've got."

So I went out and ran another mile an hour faster. That was 2 ½ miles an hour faster than anything I ran with Mario's motor. We won the pole with that motor, put one of ours in for the 125-mile qualifying race, then went back to the Jenkins/Tryson motor for the 500.

Everybody told me it would blow up in the race. They said the constant motion of the car on the track would put too much strain on a drag-racing motor and it would blow. I risked it because it was such a good motor, but, sure enough, it blew.

We had pitted early in the race and had run about 10 laps when I came off turn three and the engine quit. I yelled to Mario that I was out of gas. He yelled back that I couldn't be because we had just topped off. So I went in and we put gas in, but the engine died on pit road and wouldn't crank again.

We completed 36 laps, finished 28th and Tryson was sick about it. What happened was that when Joe built the fuel pump, he used a hollow push rod like he would in a drag-racing engine instead of a solid one like we used in our engines, and it broke.

Joe felt especially bad because he had a whole box of solid push rods on a shelf in his shop, but he'd used a hollow one out of habit.

For me, it was devastating disappointment at Daytona for a second straight year because that engine was good enough to have won the race had it held together.

Remember what I said earlier about Mario Rossi? A perfect example of his shiftiness was what happened the day after the 1975 Daytona 500.

The only stipulation "Grumpy" Jenkins put on me when he built that motor was that no one other than NASCAR could see his intake manifold or his carburetor. He didn't want any of his competitors to see it, and that was fine by me so we shook on it.

The next morning after the 500, I went into the motor room at the shop, and that engine was sitting upside down in a rack with the intake manifold and carburetor missing. I immediately said, "Where's Mario?"

Some of the guys said he had been there just a few minutes before, but he was gone. I asked the girl up front, and she said he had just left to go somewhere, but she didn't know where. I jumped in my car and hauled ass to "Smokey's."

You couldn't just walk straight into "Smokey's" garage. You pushed a buzzer, and he had a guy named Ralph who worked for him that usually came to the door. Well, Ralph cracked the door, and I snatched it open and rushed in.

Just like I figured, "Smokey" and Mario were there with "Grumpy" Jenkins' intake manifold and carburetor laying on a bench in front of them. "Smokey" didn't talk very nice, and he asked me what the blank I was doing barging in like that.

I told him "Grumpy" had built me that motor with the stipulation that no one saw his intake manifold or his carburetor, and I had given him my word on it. He turned and told Mario to get the hell out.

Mario was a real dark-skinned person, but he turned white as a ghost when "Smokey" said that. He gathered all the stuff up and started out. I was about to follow when "Smokey" grabbed me by the arm and said he wanted to talk to me. Well, I almost soiled my pants because "Smokey" was an intimidating, powerful man, and I thought he was mad because I had barged in uninvited.

But he said, "There's not a lot of people who would do what you did. I appreciate that." I told him that if he had built me an engine and didn't want anybody to see it, then nobody would see it if I had given my word.

From that day on, I had no better friend in racing than "Smokey" Yunick. Anything I needed or wanted, he'd let me have. He and his wife, Margie, used to want us to stay at their house in a garage apartment they had out back. I don't think Pat and I ever did, but he offered. We did eat dinner with them several times when we were in Daytona.

I think "Smokey" liked me because I was honest and straightforward and wasn't like Mario, always trying to make points by kissing his ass.

After the 1975 Daytona disappointment, things didn't get any better. There were two more 28th-place finishes, a 35th at Michigan and a 42nd at Talladega, which really stung since that was my home track.

Still, I was floored when Bill fired me on July 4, 1975. That morning, I had started on the pole for the Firecracker 400 and finished fifth. But standing on the back of his yacht that evening in the Daytona marina, he told me I was too old and couldn't drive anymore. Just like that, I was out and Darrell was in.

Being the president of the company didn't count for much, either. After they fired me, I got a letter from one of Bill's lawyers offering me $250 for all my stock. That stock represented two racecars, all the equipment I had like wire welders and engines, the drive-on truck, everything I owned. And they had offered me $250 for it.

This infuriated me so much I asked Bobby to put me in touch with a high powered-lawyer he knew in Birmingham. He took one look at the offer and told me the best thing I could do was take the $250 because the stock really wasn't worth anything.

I got mad, said several ugly words, then told Pat to write them back saying we'd take $500. They sent me that check back super express because they were buying everything I owned for $500.

Before I moved to Daytona in July 1974, I had a house in Hueytown, a place on the river and all my racing equipment. When I moved back to Hueytown in July of 1975, I had $2,000 in a savings account, four kids to put in Catholic school, no house, no racecars, no tools. I didn't even own a crescent wrench.

Some good friends of ours, Horace and Nell Gray, let us live with them for six months and wouldn't even let me pay rent. Bobby rented me space in a corner of his shop for $100 a month, and I found a guy that wanted to sell a Nova—one I would later give to my nephew Davey to get his racing career started—for $8,700.

I went to Sam Neilson, the main man at First Western Bank and told him I needed to borrow some money. He asked me how much I needed and how I would pay it back. I told him, "Sam, I need $9,000, and I don't know how I'm going to pay it back."

He wrote me a 90-day note for $9,000, and he told me to come back after the 90 days and he'd renew it. I bought that Nova, and with it, I paid that loan back before the second 90 days were up.

My first Friday night back at Birmingham International Raceway, I ran second and third in the features. The next week they had three features, and I won all three. That made me the only cat ever to win all three features on the same night at BIR.

And Gardner said I couldn't drive anymore.

During Speedweeks 2005, I was walking through the pits in Daytona when I saw Bill Gardner talking to Robert Yates. It was the first time I'd seen him in years, and I didn't recognize him at first. Then it came to me who it was, and I really had to hold myself back.

When I saw him, I was shocked. When he saw me, he was like, "How you doing? How's it going?" You know—buddy-buddy kind of stuff.

I almost said I had finally recovered from where he tried to put me in the ground, but I ended up just saying, "I'm fine." It was very awkward because I had a lot more to say, but didn't, which is definitely unusual for me.

I just didn't think that was the place to make a scene. There were a lot of people there who didn't know anything about what went on and wouldn't have cared. I could see us two old farts fighting right there if I had told him how much of an SOB I thought he was, and is to this day, and that would not have been good. So, hard as it was, I kept my mouth shut.

Two Trips to the White House

I may not have been the smartest person when it came to politics, but I was smart enough to know how important an invitation to the White House was.

I actually got to meet two sitting presidents in the White House, Richard Nixon late in his administration and Jimmy Carter early in his, and the two visits were very different.

Nixon spoke to us and was very nice and cordial, but he did not mingle with us. It was like he really didn't know what to say or didn't have anything to ask us about racing, so he kept his distance. President Carter made us feel like family.

I had first met him when he was governor of Georgia, and he came to some races at Atlanta. He came down to the garage and met all of us. He talked to me and Bobby about the Alabama Gang and other things that were more personal than just a celebrity to a politician.

I was interested in his thoughts about the farming industry and about him coming from a peanut-farming family. We talked about our love of the outdoors. We never did talk about his brother, Billy, though.

We talked about racing a lot, too, and he was pretty knowledgeable about our sport.

I remember he asked me one time why it was that soon after we started a race at Atlanta, you'd see Richard Petty running around the top of the track when nobody else would?

I told him that, in my opinion, Richard ran the top of the racetrack because he was scared. "Scared of what?" he asked. I said, "He's scared something's going to happen, and he knows if he's up close to the wall, he won't have to go that far before he hits it."

He just laughed, and I remember it was a very quiet laugh.

Then I told him that, to be perfectly honest, it was not a bad place to be.

At that time, Atlanta was one of the hardest tracks we raced on because it was such a tough track for handling. Your car really had to handle good there. And Richard found that his cars handled better near the top.

I told then-Governor Carter that I used to ask myself that same question when I first started out. What was Richard doing up there? Finally, I decided to go up there and find out. What I learned was that while all the other guys were slipping and sliding around the bottom, you could just keep right on trucking at the top.

Richard didn't like me finding out. He said to me several times, "What are you doing up there where I run?" I told him, "I went up there because I wanted to know what you were doing, and now I do."

Well, Carter understood all that like he was one of the guys.

On one of those trips to the garage, when he was campaigning for president, Carter told us that if he got elected, he would

Bobby, Judy, Pat and I met Jimmy and Rosalynn Carter several times over the years. On this occasion we visited them while he was the governor of Georgia. After he was elected President of the United States, they invited us to the White House. (***Courtesy of Donnie and Pat Allison***)

invite us to the White House. I thought, big deal, another campaign promise that won't be kept. But sure enough, not long after he was sworn in, we got a call from NASCAR saying he had invited us to visit.

Pat and I went, and Bobby and Judy. Richard and Linda Petty were there, and I don't remember who else. I do remember we had a super time. They took us through every bit of the White House, and it wasn't a deal where we had security guards breathing down our neck at every turn.

We went through the security at the main gate like everybody else did, but once we got inside, we walked around anywhere we wanted and went in any room we wanted. Everybody we met was so cordial; you'd have thought we were dignitaries from another country instead of racecar drivers from the country.

We had lunch on the White House lawn, and the entertainment was Willie Nelson, which is kind of funny considering all the tax trouble he had with the government later on. But we ate and visited with President Carter and Rosalyn and talked like we were at a Sunday social.

I remember being impressed that, throughout the whole affair, he never once mentioned that he had said he was going to invite us or anything like that. I also came away impressed because a politician actually kept one of his campaign promises.

I think President Carter had a lot of good ideas about a lot of things at the time, but like all politicians, he never got to implement most of them. He's a good man who treated us with the utmost respect and courtesy, and I'll never forget it.

Hot Time in Hot'lanta

One of the most memorable of my 10 career victories was the one NASCAR tried to take away from me and give to Richard Petty.

It happened on Nov. 5, 1978, at the Atlanta International Raceway and, believe me, things hadn't been that hot in Georgia since Sherman left town.

First, a fire in the parking lot made it look like somebody had dropped an atomic bomb there and caused a 28-lap caution while the safety crews went to fight it. Clouds of black smoke rolled into the sky, and about 30 cars were damaged or destroyed after somebody's catalytic converter set the grass on fire.

Then a mix-up by NASCAR's official scorers set my temper ablaze.

Early in the Dixie 500, I made a mistake. On a green-flag pit stop, I came off the banking in turn four too fast. The apron coming up from pit road was real steep, and when I hit it with my left front, it bashed the tire so hard it caught the fender. That bent the fender all up and hung it up in the tire.

When I finally stopped in the pits, it took two laps to get the fender untangled.

All I could say to myself after I got back out there was "Don't quit. Don't quit like a lot of people would if they got behind and just ride out the rest of the race." I knew I had a good car, and I was just that way.

Buddy Baker had the fastest car, the black-and-silver No. 28 that belonged to Harry Ranier. Late in the race, I was following Baker off turn two when his engine blew. When he blew, he waved his hand at me, and I went to the outside of him, and when I came around, the caution light was on.

I made up one lap there, but was still a lap down, so I was at the front of the inside line. When the re-start came, I beat everybody down into the first turn, and when I came down the back straightaway, the caution light was on again.

Richard Childress was still driving then, and he had spun out on the re-start. So I had just made up my two laps in only six laps of racing.

From that point, there was no question in my mind I was going to win. Before the last re-start, I even told "Hoss," "You better get NASCAR ready because I'm fixing to win this race."

Dave Marcis was leading, I was second and Petty was third on the last re-start. Coming off turn two on the last lap before the green, Petty tried to come by me on the outside, and I pinched him against the wall and motioned for him to get back. Even Dale Inman, his crew chief and cousin, told him on the radio I was on the same lap.

You can imagine my surprise when I got to Victory Lane after winning the Dixie 500 at Atlanta Motor Speedway on November 5, 1978, and was told that they were expecting Richard Petty because at first he was credited with the win. **(Photo by Himes Museum of Racing Nostalgia/Courtesy of Donnie and Pat Allison)**

Marcis maintained I shouldn't have been on the same lap with him and Richard. He said he wasn't worried about me. He said he didn't race me down in the corner.

Well, he didn't race me down in the corner because he couldn't race me down in the corner. My car was too fast. He couldn't race me all day.

I won the race and went to victory lane and Petty, who had nipped Marcis in a photo finish at the line and thought he'd ended a 42-race winless streak, went to victory lane, too. I went to the press box, and Petty went to the press box, too. It was a circus.

After it was over, I went down into the garage, and Bill France Jr. and all the NASCAR officials were gathered in a little sign-in shack for the garage area. He told me to go in and let Joe Epton, the official scorer, show me how I didn't win the race.

I told him I'd go in, but Epton couldn't show me how I didn't win the race because I won it. I went in, and I'll never forget it because Joe Epton was sitting there sweating like he was in a sauna bath when it actually was pretty cold outside.

We went over the scorecards, and I showed him where I made the laps up. He really didn't know what to say.

While I was there, I heard somebody talking in a little room in the back. I opened the door, and it was Maurice Petty, Dale Inman and Bill Gazaway.

I pointed my finger at Maurice Petty and said, "You'd take this race, too, you son-of-a-gun even though you know you didn't win it." Then I shut the door. I didn't wait for anybody to say anything to me.

After I finished with Joe Epton, I went outside, and Bill France Jr. was standing there. I walked up to him, poked him in the chest and said something like, "Don't take this race away from me. If you do, I'll whip your butt until your momma don't know who your are, and I'll never run another NASCAR race. I won this race."

I was fit to be tied. When I finally left the track that night, they had still not announced a winner. The next morning Clyde Bolton, the long-time NASCAR beat writer for The Birmingham News, called and said, "Donnie, they gave you the race." I said, "Hell no, they didn't give it to me. I won it."

And I did, but to this day, nobody from NASCAR has ever called to tell me that. Ever.

Bill France Jr. was on TV years later talking about this very thing, though, and he told a story about his son, Brian, I found very interesting.

Brian, who was 16 at the time, was in the press box that day, and he told his daddy and momma I won the race. Naturally, Billy told Brian to be quiet. The next thing I knew, they came and got Billy and told him he better check on Brian because Brian was having a press conference up in the press box to say I won.

Billy said on TV that Brian was pretty good even then because he knew who won the race, and he wasn't going to quit until the right winner got paid. That's why Brian France and I are good friends today.

Even though he's a relatively young man in the racing business, I respect Brian France greatly. The morals he has are Allison morals. No matter how much it may hurt at the time, do what's right. My parents raised me like that.

While "Hoss" and I posed for pictures in Victory Lane after the Dixie 500 fiasco, we still had not officially been credited with the win, and many Richard Petty fans in the crowd were letting us hear about it, too. (**Photo by Himes Museum of Racing Nostalgia/Courtesy of Donnie and Pat Allison**)

Publicity-wise, it would have been much more appealing for Richard Petty to have won that race than for Donnie Allison to have won it. And I definitely feel some people tired to take it away from me.

Everybody thought at the time, "Hurray, Richard has finally won another race. Hurray, Richard broke his long losing streak." But it wasn't right.

I won that race, but it didn't show it on the scorecards at first. I was told later by some people in the scoring tower the reason my scorecard got screwed up was because my scorer stopped scoring me. They also said that at the end of the race she was pulling for Petty to beat Marcis, too.

However it got screwed up, the right thing to do was fix the error and declare me the winner, no matter how unpopular it might have been. Brian France was smart enough to see that, and he was just a kid.

That's why when word came some years back that Brian was going to succeed his father as CEO of NASCAR and a lot of folks said, "Oh no, they're in trouble," I disagreed. I told them NASCAR was not in trouble; it was in very good hands.

And in the time Brian's been at the helm, he's proven me to be right on the money.

"Make Them Show You the Tape"

The funny thing about the fiasco with Cale in the infield at Daytona was that nobody could believe it was Bobby fighting and not me. When I got back to the garage, even my mother scolded me, and I had to tell her it was not me.

Nothing was funny about what NASCAR did, though. They fined me, Bobby and Cale $6,000 each, and they put Bobby and me on probation.

In those days, $6,000 was a lot of money, so that stung, even though NASCAR told us they'd refund $5,000 to us—$1,000 per race—as long as we stayed out of trouble. But what stung us as much was being put on probation when Cale wasn't.

The way we saw it, he was guilty of starting both the wreck and the fight. So we appealed to the National Stock Car Racing Commission, and a hearing was set at a Red Carpet Inn in Atlanta in early March.

AS I RECALL...

On raceday, I had told Bill France Jr. my side of what happened. How Cale had hit me in the back first, before we ever hit side to side, the whole nine yards. Now we were going to look at films to see if I could prove it.

The wreck with Cale Yarborough took us both out of contention on the last lap of the 1979 Daytona 500 and took away the best shot I ever had to win "The Great American Race." **(Courtesy of Donnie and Pat Allison)**

The day before we were supposed to go to Atlanta, I got a strange phone call. Somebody who wouldn't identify himself said, "Donnie, they have a film that shows the wreck. Make them show it to you."

I tried to find out who it was, but he wouldn't tell me. He just said again, "They have a film that shows the wreck, make them show it to you," and he hung up.

To this day, I do not know who it was that called me. I thought I did at one time, but it turned out I didn't.

We got to Atlanta, and they had a room and video equipment set up. There was Bill Gazaway, NASCAR's competition director, Les Richter from Riverside, Calif., and John Riddle from Dover and two other people that made up the appeal board.

The first thing Gazaway said to us was he wanted to show us the films one at a time. Bobby said, "Of what?" and Gazaway said, "The wreck."

Bobby quickly told him he was appealing being suspended for the fight, not the wreck. So they asked him to step across the hall so they could show me the films.

Gazaway said they had five films, and they wanted me to watch them in their entirety, then I could comment.

The first one was the main CBS film, and all I could see was crashing cars. I couldn't really tell what was going on. So they put the second one on, and it was another CBS film shot from more of an angle in turn four. I could see more of the cars then, but it still wasn't conclusive.

The third film was the one I was looking for. It came from a TV station in Jacksonville, Fla., and it was filmed from somewhere in turns one and two. Bottom line, it showed Cale's front bumper hitting my left-rear bumper while we were still in the racetrack.

When it was done, I asked Gazaway if he could run it in reverse until I told him to stop. When I said stop, it went a few frames past what I was looking for. I asked if he could run it forward in slow motion until I told him to stop again. He said he could run it forward frame by frame.

If I live to be 200 years old, I'll never forget what happened when I said stop again. Les Richter stood up and said, "Why in the hell haven't we seen this film?" End of appeal.

Bobby never had to watch anything, and they put Cale on probation, too. They said he appealed, but we never heard another word about it, ever.

What a lot of people may not remember is that the next week at Rockingham, Cale and I wrecked again on the ninth lap. It was a very bad situation. I tried not to hit him, and as a result, we crashed harder and took a lot of cars out of the race, including mine.

I saw Bill France Jr. after that accident, and I told him, "Really, it's your fault. You put me on probation, and I'm trying not to have any problems. I should have just knocked Cale out of the way and went on."

He just looked at me and grinned.

To be honest, though, I believe explaining my side of it the way I did to Bill Jr. after the wreck that day in Daytona was what got that film sent to Atlanta. I'm 99 percent sure it was Bill France Jr. that had that film sent to Atlanta, and I'd say the phone call came from someone close to him, too.

For a time, I thought it was Jackie Arute, the guy you see now covering football for ABC, who made that phone call. He was a racing promoter from up in New England, and he was working for NASCAR in some capacity then.

But Jackie has sworn to me it was not him. I've asked him and he said, "Nope, it wasn't me." I think he knows who it was, but he still won't tell me.

Whoever it was, thanks for dropping a dime.

"The Intimidator"

Dale Earnhardt and I were good friends from the time I threatened to whip his ass at Charlotte early in his career until the day we lost him.

He was behind me coming out of turn four that day in practice, and he ran into me pretty hard going down the front straightaway. When I got into turn one, I looked in the mirror, gauged it just right and put the brakes on. When he got against me, it virtually stopped him. He pulled down beside me, and I shook my finger at him.

We went to the garage where I was parked on one side and he was on the other. I jumped out of my car because I was pretty aggravated and ran around to where his car was.

By then, my crew knew something was wrong, but they didn't know what because I hadn't told them what happened on the racetrack. Apparently he had told one of his crew members because when I got around there, the crewman saw me, and his eyes got as big as saucers.

Dale turned around and started to say something, but I stopped him. I told him, "Don't open your mouth. Don't say one word. If you ever run into me in practice again, I'm going to wreck you right then and then get out and whip your ass."

My crew was there by that time, and they were all wondering what was happening, but I didn't say any more. I had said my piece to Dale, and didn't need to say anything to anybody else.

After practice, he came down and asked if I had calmed down enough to talk. I told him I had been calm the whole time because I hadn't done anything wrong to begin with.

Dale told me again he wanted to talk, and I told him again there was nothing to talk about. I said I had already told him not to ever run over me in practice again, or I would wreck him, and I meant it. I told him there was no reason to run over anybody in practice.

He said he knew that, and admitted he had misjudged and gotten too close to me. I told him, "Fine. Now you go your way and I'll go mine." It never came up again, and I never had a problem with Dale on the track again.

It was obvious from the start Dale was something special. On the track, Dale Earnhardt was the ultimate racecar driver, just like my brother Bobby. Every time he got in a racecar, he gave 150 percent. No headaches. No feeling bad. No arguing with the wife that he took to work with him.

When he got in that racecar, there was nothing else. Just the race. That's how he won all those championships. He stayed after it all the time.

Many people had a great misconception about what the man was off the track, however. A lot of that had to do with the media and with Dale himself.

He wasn't that well educated, and he learned very early on to really watch what he said for fear of being misconstrued, and rightly so. From my experience with the media, you could tell four guys something in an interview and be lucky if half of them wrote the same thing in their stories.

But if you knew Dale behind the scenes, if you were his friend, you could say anything you wanted to him, anytime you wanted to say it.

There's a story about Dale and his generosity that will stick with me forever because it really showed what kind of man he was.

There was an automobile dealer in Birmingham, Ala. named Jim Burke who told me one time he'd give anything if Dale would come to his dealership and sign autographs. He knew I knew Dale, and he asked if I could help make that happen.

I told him I probably couldn't, but my wife could. Dale used to say I was his friend, but he was really Pat's friend. Well, I told Dale what I'd like for him to do and, sure enough, he told me to tell Pat to come talk to him.

We were at Darlington, and Pat went right up into the truck and told Dale she wanted him to come to Birmingham for Neil Bonnett Appreciation Day. He said, "You know I get $25,000 per autograph session, but here's what I'm going to do."

He told Pat to have Jim write Susan Bonnett, Neil's widow, a check for $10,000 and another check for $5,000 to the Head Injury Foundation, and he'd do it. When we told Jim, he was so overjoyed he said he'd give whatever Dale wanted to whoever Dale wanted.

The day came, and Pat went and got Earnhardt at the airport. That was one of his specific instructions. Pat had to be the one to pick him up at the airport after he flew down in his private plane.

AS I RECALL...

It was me, Dale, Stanley Smith, Mickey Gibbs and a few other Alabama drivers set up to sign autographs at Jim's dealership. The session was supposed to start at 6 p.m., but long before that, people were lined up from the front door all the way across the interstate that ran in front of the dealership.

About 10 minutes before we were actually supposed to start, Dale looked out at the crowd and said, "Well, there's a lot of people here; let's get started." And we signed autographs for a line of people that never got shorter.

We were supposed to stop at 8 p.m., and they came and told us they were going to put a uniformed policeman in line so they could cut it off. But Dale told them to wait a little bit and he'd let them know when to do that.

We signed past the time we were supposed to stop, and finally Dale turned around and told somebody to get the policeman for him. I know he hated to leave people in line, but it was the only way we were going to get home before sunrise.

After that, we went into a hospitality room they had there for the big wheels at the dealership, the drivers and their wives. Jim walked over and thanked Dale for coming and told him how indebted he was to him for it.

That's when Dale did something I'll never forget. Dale Earnhardt looked straight at Jim Burke and said, "Nope, I'm indebted. To be able to do this for Susan Bonnett and my friend, Pat, makes me feel so good. Thank you."

That's the side of Dale most people never knew. He didn't have to do any of it. But he flew his own plane there, brought his own people, and he did it because he wanted to help out Susan Bonnett and Pat.

I wasn't at Daytona International Speedway that day in 2001 when Dale died. I had been in Daytona that morning, but Pat and I did not go to the race. We left early to get back to Salisbury.

We were somewhere around Greenville, S.C., when word came that Dale had crashed. I was listening closely, and the thing that warned me something bad had happened was when they said that Ken Schrader ran over to the car, then turned right away and began motioning for the safety crew to get there.

I told Pat, "That's not good."

We came on home and went by my daughter Pam's to pick up our little dog, and we met my son-in-law, Hut Stricklin, coming out of the driveway. He told us Dale had died, and he said we wouldn't believe it when we saw it because the wreck didn't look that bad.

It's the same thing millions of others said as well, and I'll admit, I was shocked when I saw it. You would never imagine in all your life a wreck that didn't look any worse than that would have killed "The Intimidator" Dale Earnhardt.

From all appearances, it wasn't that hard a hit. Even Schrader helped turn him a little from going straight into the wall. Dale was turning to the right, but Schrader hit him and straightened him some, which should have cushioned the blow a little bit. But it didn't.

I've said it a thousand times since then. It was no different than my nephew, Clifford, getting killed in a wreck at Michigan in 1992 or others who have died in racecars. We may not understand it, but there's a reason for it. What that is, I can't say. But there is a reason.

Old Jabber "Jaws"

Many of the drivers of my era—especially Bobby—might not agree with me on this, but I always thought Darrell Waltrip was good for our sport.

Until Darrell arrived in 1972, we had never had anyone do what he did. That was, be a Muhammad Ali. Somebody who said what he was going to do and then did it.

Dale Earnhardt was a little bit like that, but the difference was he didn't say it. He just did it. With Darrell, you heard everything there was to hear about it and then some.

That's the thing that hacked so many people off with Darrell. He had a really big mouth. The older guys like Richard, Cale and Bobby didn't care for Darrell because he'd run his mouth so much, but then he'd go out and back it up. That's what really pissed a lot of people off with him.

The thing that upset a lot of the fans with Darrell was that he was so brash. He got on the top dogs like Richard, Cale and Bobby. He had run-ins with them, he came out on top, and that made their fans mad.

I didn't like the way he ran his mouth, naturally, but I never had the problems with Darrell others had. In fact, I believe his emergence as a star was very important for NASCAR.

After the factories got out of racing in 1970 and 1971, things stood still for a while. It was kind of quiet. Then all of a sudden, here came this young kid from Owensboro, Ky., via Nashville, and not only did he have an old Mercury his father-in-law was helping him race, he had a loud mouth.

Whether we liked it or not at the time, it wasn't too long after he showed up that he was one of the main topics of conversation when it came to NASCAR racing. And that was good because it meant people were paying attention to us again.

I was there that day at Darlington when Cale and some others hung a six-foot rubber shark in Darrell's garage and nicknamed him "Jaws." So what did Darrell do? He made the best of it.

I'm sure he didn't like it, but he made a big deal out of it. "Yeah, I'm 'Jaws,'" he said. Whatever you wanted to call him was fine, as long as you called him a winner, too.

The person I always felt bad for through all of it, though, was Darrell's wife, Stevie.

I was pretty fortunate to start up front at lot of times in those days, so I stood there through a lot of driver introductions. Every time the man got introduced, everybody in the stands would boo.

I would turn around and look at Stevie sometimes and wonder what she was feeling. Can you imagine standing there the wife of a man who's doing pretty good in racing, but every time he was introduced it brought the house down with boos? It had to be a terrible feeling for her.

AS I RECALL...

The truth was that Darrell was stepping on Cale Yarborough's toes, he was stepping on Bobby Allison's toes, even Richard Petty's toes a little bit, although Richard was on the downhill slide by that time anyway. He wasn't over the hill by any stretch, but if Bobby and Cale ran good, they were ahead of Richard. And Darrell was often ahead of them.

A lot of people said a lot of derogatory things about Darrell back then, but you don't win 84 races—which ties Bobby for fourth most in NASCAR history—if you can't drive. You don't win three championships if you can't drive.

Looking back on it, nobody should have booed Darrell for what he did. If you liked competition, you had to like Darrell Waltrip.

Now Darrell is one of the analysts for the Fox broadcasts, and he's good at that, too. He sometimes goes a little overboard talking about what us old guys used to do, but then Darrell was always one to toot his own horn.

One thing I will give him credit for is that he does a good job of explaining racing to the casual race fan. It's hard sometimes for somebody like me or Darrell or any driver who's been there, done that, to explain racing in simple terms.

He has to remember he's not talking to me or his crew chief or somebody who's going to understand all the technical stuff. He's talking to the public and has to explain it in a way the public can understand. Old "Jaws" is definitely good at it, but then talking always was one of his strongest suits.

I think I would have been good at it if I could have gotten the chance to do it full-time, too. I did call some races over the years for ABC and ESPN, and got to work with guys like Jim McKay and Keith Jackson in the process. In fact, I was in the booth with Keith when Bobby won his last race in Daytona, and it turned out to be the last race Keith ever called for ABC as well.

It was the Firecracker 400 on July 4, 1987, and it was a scoring fiasco that ended with ABC not getting a shot of Bobby taking the checkered flag. But that wasn't the network's fault. It was the folks scoring the race in NASCAR control that screwed that one up.

Bobby had been down one lap for most of the race while Buddy Baker and Ken Schrader battled it out for the lead up front. But when Rick Wilson hit the wall with nine laps to go, Bobby got his lap back, and when the race re-started with five laps to go, he was in 13th place.

I had been telling Keith during the breaks that if Bobby got back on the lead lap, he would win the race because he had the fastest car. Sure enough with just over a lap to go, Bobby swept past Baker, Schrader and Dave Marcis down the backstretch and took the lead.

But somehow they had missed it in scoring, so nobody else seemed to realize that Bobby was in the lead. I kept telling Keith he was leading, and he kept asking the girls they had running down to our booth with pieces of paper that had the positions on them if Allison was in the lead.

I kept saying he was, but NASCAR never confirmed it, so he wasn't sure. Bobby was leading the pack when they came off the fourth turn on the last lap, and then Schrader went out of control, flipped upside down and got T-boned by Harry Gant. Bobby beat Baker to the line by four seconds, but the ABC cameras were focused on Schrader and didn't catch it.

Keith had to catch up by saying something like, "There goes your winner Bobby Allison down the backstretch."

After the race was over and we went off the air, he blew up. He was like another person entirely. He cussed out the people in scoring. He ranted and raved and told all the ABC people there that he would never do another race and, to my knowledge, he didn't.

Keith was so upset because he was a perfectionist who wanted things done right, and I loved working with him. He was always incredibly prepared. I remember going to his hotel room the night before we did our first race together, and he had all this information spread out on his bed, and he made tons of notes.

He posted a lineup on the window in front of us, little paper tags he made that had the driver, the car number and the starting position on them. If somebody fell out, he would take the tag and move it from one side to the other so we knew who was who.

He was thorough, he knew what he was talking about, and he was just fun to talk to. He was a Georgia boy and loved to talk about college football. He liked Georgia, of course, and the Southeastern Conference. He liked Alabama, he was close to Paul "Bear" Bryant, and he liked to talk about both.

I always loved to listen to him call college football games more than anybody else going. Alabama-Auburn, Southern Cal-UCLA, The Rose Bowl, all the games Keith called over the years ... there ain't been nobody better. He was just a No. 1 class guy all the way around, and I'm one of the lucky ones who can say they've been privileged to share a microphone with him.

I also did *Wide World of Sports* several times, and I got to hear Jim McKay say, "Ladies and gentleman, Donnie Allison's *Wide World of Sports*." That was a real rush, too.

When the new television deals with Fox and NBC went into effect in 2001, I was interested in doing some more broadcasting.

In fact, I said something to Mike Helton about it. I asked him if NASCAR could put in a good word for me with the new networks about getting one of the analyst jobs. But Fox had already decided on Darrell, Jeff Hammond and Larry McReynolds, and NBC was set on Benny Parsons.

All those guys did a pretty good job of talking things—including themselves—up over the years, and it paid off when the TV folks came calling.

I never talked much about my racing, although I had a lot of reasons to toot my horn, too. But it wasn't my nature. I just didn't believe I had to stand around and talk about winning the race or how good I was. I honestly believed that when the checkered flag fell on Sunday, that told everybody how good or bad you were.

Besides, I wasn't trying to please anybody else. I tried to please myself, and I felt if I pleased myself, everybody else had to be pleased.

Looking back, I should have talked more. Then maybe more people would know that I was a damn good driver. That I drove hard and smart. That when I was in a good car, I ran good in every division I ever raced in.

Yes, I should have talked more. Or at least had a better P.R. person.

Saddling Up with "Hoss"

A.J. Foyt was a jerk to me plenty of times over the years, but the man helped me plenty in my racing career, too. Besides putting me in an Indy car, A.J. was—indirectly at least—responsible for me getting the best NASCAR ride I ever had with "Hoss" Ellington.

A.J. and "Hoss" had teamed up to run a partial Winston Cup schedule in 1976, but A.J. still planned to spend the month of May at The Brickyard going for another Indy 500.

He had planned to fly back and forth and run at Talladega with "Hoss" during that month, too. But when a rainout at Indy caused a conflict for A.J. at Talladega, "Hoss" called me.

I had driven two races for "Hoss" in 1975 after being dumped by DiGard, including a third-place finish in the Talladega 500 in August. So "Hoss" figured I was a good fit for his red-and-white No. 28 Chevy, and he was right.

I qualified third and finished seventh in the Winston 500, and that car could have won the race if not for some bad luck right out of the chute. We started the race with the valve covers leaking oil, so I had to come in, and went down three laps changing those.

Winning the National 500 at Charlotte in October 1976 was especially sweet because it was my first win in five years and because it earned me a full-time ride with "Hoss" Ellington. **(Courtesy of Donnie and Pat Allison)**

Once we got that fixed, I could run with anybody, including Buddy Baker. Buddy won the race that day with an average speed of 169.887 miles per hour, making it the fastest 500-mile NASCAR race ever run to that point.

The frustrating thing was, once we got the problem fixed, I actually ran Buddy down and passed him to get one of my laps back. But I was so far back, I didn't have any shot at winning. Still, seventh was good enough for "Hoss" to keep me in the seat the rest of the month.

Two weeks later, I drove for him again at Dover Downs and finished 35th, but two weeks after that, on May 30, 1976, I started seventh and finished sixth in the World 600 at Charlotte. That same day A.J. finished second to Johnny Rutherford at Indy.

A.J. returned to run the Firecracker 400 in Daytona on July 4 and the Talladega 500 in early August. Then "Hoss" called me again to run the car in the Southern 500 at Darlington on Sept. 5, 1976. Well, I could have won that race, too, if bad luck hadn't struck again.

I was leading the race on lap 167 and just about to lap Pearson in the Wood Brothers car when the water pump busted. Pearson went on to win the race, and I went home disappointed again.

Then came the National 500 at the Charlotte Motor Speedway on Oct. 10, 1976.

In addition to entering a car for A.J., "Hoss" decided to enter me in a Monte Carlo he had prepared, but that A.J. wouldn't drive. Driving that Monte Carlo was a sore subject with A.J.

He and "Hoss" had a definite difference of opinion on the Monte Carlos. "Hoss" liked them, A.J. didn't. A.J. told "Hoss" he was just copying "Junior" Johnson, who had switched to Monte Carlos first. A.J. said it was a monkey-see, monkey-do way of building racecars.

So if I was a betting man, I'd bet it made A.J. pretty mad that day when I passed him for fourth place fairly early in the race driving that Monte Carlo he hated so much.

Finally, after 59 laps, A.J. pulled into the pits, got out, said some nasty things about his car not being prepared properly and told "Hoss" he would never drive for him again. Then A.J. headed out to find a plane to Houston.

Meanwhile, I was enjoying the view from the front. It had been five years and a bunch of miles of bad road since I'd last gone to victory lane, and I was loving the trip. Afterwards, "Hoss" told A.J. and me that I would be his full-time driver from then on.

It was just another case where A.J.'s ego and his temper got the best of him. He didn't want that Monte Carlo. The way he felt was that he didn't want to have to copy "Junior" Johnson to win. That's why he said what he said.

But if A.J. Foyt had gotten in that Monte Carlo that day, I probably never would have driven full-time for "Hoss" like I did.

There's no doubt "Hoss" and his crew were the best team I ever drove for.

Shelton "Runt" Pittman was the crew chief, team manager or whatever you want to call him, and he never got the credit for running the show he should have until later in his career. We had a nucleus of people—Pittman, Jackie Rogers, Norman Miller— that was unbelievable.

Nothing was too hard or too much for them to do. Anything that made the car run better, they did it. That's why we posted three wins, 22 top-five finishes and seven other top-10 finishes in the 60 races I ran for "Hoss" over the next four seasons and change.

I started on the pole for the 1977 Daytona 500, and that started a string that saw us qualify top 10 for all 17 races we ran that season. I won at Talladega in August and again at Rockingham in late October.

I won for the last time in my career at Atlanta in November 1978 and, as I've already said, I should have won the 1979 Daytona 500. Things went downhill fast between "Hoss" and me after that race, though.

My crew chief, Shelton "Runt" Pittman and I talk over our strategy. Working with "Runt" and "Hoss" was one of the best experiences of my NASCAR career. (**Courtesy of Donnie and Pat Allison**)

For one thing, after all that had happened on the track between Cale and me that day, "Hoss" still ended up giving Cale a ride back to North Carolina in his van. Cale couldn't fly his airplane home because everything was snowed in, so he hitched a ride with "Hoss."

Believe me, I had a few ugly things to say about that. Here the guy had just knocked us out of winning the Daytona 500, and "Hoss" gave him a ride. I asked "Hoss" why the hell he did that. "Hoss" said he didn't see that it hurt anything, and that's just the way he was—a good guy.

I told him what I felt, though. I told him I didn't think that was the thing to do, and I didn't like it. I don't know whether "Hoss" changed his mind after that or I changed mine, but things definitely changed between us.

I ran pretty good in some races after that, but for some reason, it just never was the same. I won a pole at Darlington in early April and finished second at Michigan in June and Dover in September, but the chemistry was gone.

I started the 1980 season pretty good, starting on the outside pole and finishing seventh in the Daytona 500. Then I finished fifth at Rockingham in the fourth race and started fourth at Atlanta in the fifth. I fell out of that race because something went wrong in the rear end and finished 26th.

After that race, "Hoss" and I were finished. I don't know if it had anything to do with that rear end going bad. I don't remember exactly what it was. All I remember is that "Hoss" and I got into a heated discussion one day, and I told him he needed to get someone else to drive the car. So he did.

"Hoss" called Pearson, who had been sitting the season out at home in Spartanburg, S.C., and I'll be damned if two weeks later he didn't go to Darlington and win his first time out.

Now David Pearson had called me on the phone about three times in my racing life, but he called me after that race to tell me he couldn't believe how good that car drove. That really stuck in my craw because of all the races I should have won at Darlington but didn't, and then Pearson goes out the first time and wins there.

Looking back, I regret that it ended that way, and "Hoss" does, too. He tells me all the time he wishes we had kept that going. At the time, we had a disagreement we just couldn't put to bed. Back then, it just seemed like it was time to go. But it really wasn't good for either of us.

AS I RECALL...

I will admit this. After I got out of "Hoss" Ellington's car, I never got back in another good racecar. Later that year and into 1981, I drove for Kenny Childers, and his cars were decent, and that's all I can say. Just decent. After that, the rest of them were pretty bad.

Yes, if "Hoss" and I had kept things going maybe we would have won a lot more races together. And maybe I wouldn't have almost gotten killed the next year at Charlotte.

Busted Up Bad in Charlotte

John Rebhan was a good man, but he owned a bad racecar. Bad enough that I got my ass busted in it.

When I got in Rebhan's Oldsmobile for the Winston 500 at Talladega in May 1981, it had already been through one terrible crackup that season.

In the first Twin 125 qualifying race at Daytona, a guy named John Anderson was driving the car when he came off turn two and spun, then went skipping down the backstretch like a stone on a pond.

Anderson eventually hit the infield grass, flipped over backwards and landed on his roof, then flipped another five times. It was, without a doubt, one of the most spectacular of all the spectacular wrecks we'd seen at Daytona.

So the car had problems, and I knew it, but at that point I had to take a ride where I could get it, and I felt like it could be a good opportunity. Also, the crew chief was legendary racing man Harry Hyde, and I trusted Harry, which proved to be a mistake.

I ran fifth for Rebhan that day in Talladega. I say I ran fifth because the car was not a fifth-place car, but I used my knowledge of the draft and everything there to finish up front. Then we went to Dover, Del., and the car was terrible.

The next race was the World 600 at Charlotte Motor Speedway.

I went to Harry and said, "Let's don't go to Charlotte. Let's go test and get things straightened out so we can go race." Well, Harry said we had to run Charlotte because it was his own backyard, which, after I thought about it later, was his ego talking.

Harry Hyde was good with a racecar and deserved credit for it, but he was also very egotistical and couldn't admit to people in the NASCAR community he didn't have a car that would run.

At that time, everybody was trying to out-engineer everybody else, and Harry was right there with them.

We had a good motor in the car, one built by Randy Dorton, who was killed in the Hendrick Motorsports plane crash at Martinsville in 2004. But the aerodynamics were terrible.

It's very technical, but we were trying to run a rear-steer car, and you couldn't do it with that type car. What got me, though, was there was absolutely no reason for us not to go test and figure out something to make the car handle better.

The factories weren't involved. Rebhan had his company name on the side of the car, so we didn't have any sponsors to satisfy. We weren't running for the championship. Here was a guy in Harry Hyde who was supposed to be the world's greatest crew chief, and I couldn't get him to come to his senses enough to go test and get our problems straightened out.

That says to me it was Harry's ego, plain and simple.

Still, they were going to wave a green flag, and I had a ride, so I went to Charlotte despite my uneasiness. I didn't believe in premonitions or anything like that, but my experience told me something bad could happen to me in that car.

So much so that the day I left for North Carolina, I told Pat, "I'm going to mess around in this junk and get my ass busted," and that weekend, I did.

I don't remember much about May 24, 1981. My neurosurgeons told me I'd probably never remember anything about the wreck itself, so what I do know, I've pieced together over the years from media reports and conversations with friends and family.

One thing I do remember, though, is a confrontation with a security guard on pit road that morning. I had my twins, Ronald and Donald, with me, and he was trying to run them off. I told him they were with me, and he said they weren't supposed to be there. I told him they were out there, and they were going to stay.

That's the last thing I honestly remember until I woke up in a hospital bed in Birmingham, Ala. a week later.

The other thing I remember is that the car was junk. I was already about to go four laps down less than 200 laps into a 400-lap race, and you can look at my record and see I didn't get lapped almost once every 50 laps unless a wheel fell off or something.

From what I've put together, I was going into the fourth turn on lap 146 when I hit the wall and spun. I don't know why I hit the wall and spun, but I did, and I was going down into the infield when Dick Brooks blasted me on the passenger side.

The collision knocked me unconscious and badly broke my left leg and five ribs. It fractured my right shoulder blade and my cheekbone and left me with a collapsed lung and a serious concussion. Brooks walked away with a broken shoulder blade.

I was in bad shape when they loaded me on a helicopter and took me to Charlotte Memorial Hospital. When I got there, at first they were worried I had also ruptured my aorta and suffered other internal injuries—so worried they called a priest because they thought I might need last rites.

But thanks to the prayers of my family and friends, my mother wearing out two or three sets of rosary beads, and the fine work of the medical staff, the priest wasn't needed that day. Thank goodness.

One thing in particular has always bothered me about that wreck, and it's something I want to clear up. I don't remember what happened that day or why, but I know me, and I know I didn't crash from trying too hard because I was so many laps down.

The year before in the same race, I had parked a car I could pass the leader with because the pit crew wasn't any good. I kept losing two laps every time I pitted, and I didn't think that was any way to run a race, so I parked it.

Why I kept running that car that many laps down, I still haven't figured out and probably never will. Why I crashed that day, I'll never know. I know for sure, though, that the crash, for all practical purposes, ended my racing career.

The Road to Recovery

The day before I left to go to Charlotte, I had hung the garage doors on a brand new shop I had just built on my farm in Faunsdale, Ala. I had also just recently bought a new hay baler, and had a bunch of hay in the field I planned to bale when I got back.

The wreck changed all that.

Instead of happily returning to the farm to bale my hay, I spent a week going in and out of consciousness in the Charlotte hospital. Then I was transferred to Carraway Methodist Hospital in Birmingham where they operated on my broken leg, which proved to be the worst of my injuries.

A funny thing had happened while I was in the emergency room in Charlotte. At least they told me later it was funny.

Bobby, who had won the race that day, came in and started talking to me while I was sedated, and I kept saying, "Hay, hay, hay." He didn't understand what I meant so he leaned over and yelled, "What did you say?" I guess he thought I was trying to tell him I was glad to see him.

After a while he figured out that somehow, through all the pain and the fog of medicine, I was trying to tell him I needed to get that hay baled at the farm. So a good friend of mine, Jack P. Davis, took all my boys, and they went home to Alabama the next day and baled that hay.

It's humorous now, but those thoughts of working on the farm—and later actually doing it—are what brought me back from the brink.

I stayed in Birmingham for another week, and I was having a bad time because, with my head injury, I still wasn't 100 percent. But I was coherent enough to understand the day the doctor told me I could go home, but only if I agreed not to put any weight on my left leg.

He said he wanted me to promise I wouldn't even stand on it to shave because, with the type of fracture I had, even that could mess it up. I said OK, I wouldn't walk on it until it was time.

Another friend, Ray Fenton, came up from Florida to help Pat take me home to Hueytown from the hospital but, of course, I told them I wanted to go to the farm.

They said if I felt better the next day they would take me. So the next day I got up, ate and did all my stuff to get ready. Then I got on my crutches, got myself out to the car and blew the horn.

I was in the right front seat when Pat came out and said "What are you doing?" I said, "Either you take me to the farm or I'm going to take myself to the farm." So she got Ray, and we headed out.

We had gotten about halfway there when they told me the reason they hadn't taken me to the farm the day before was because they had some bad news, and they wanted to make sure I was ready for it. While I had been in the hospital, somebody had broken into the new shop and stolen all my tools and a generator I had borrowed from a friend.

So we went down that day and checked on all the damage and went back home to Hueytown. But I was having a tough time of it there.

I had a lot of get-well cards I couldn't read. Everybody around me was well. Bobby was racing. And I was sitting in my chair thinking, what am I going to do? What's in store for me? Am I ever going to race again?

Finally, one day I just said, "Donnie, we've got to quit feeling sorry for ourselves. We've got to get busy. We've got to get well."

That's when I headed to the farm and stayed there. That was in July of 1981, probably three weeks after all my surgeries, and that really started my rehab.

I hadn't been there long when I looked out the window and saw my tractor sitting in the pasture with a bush hog on it so I headed out on my crutches towards it. Pat asked me what I thought I was doing, and I told her I was going to run my tractor.

She told me I couldn't do that, and I told her to watch me. I put my crutches in the bars that supported the cab, got myself up on the tractor, put it in gear and proceeded to bush hog around in a small circle.

I cut a little bit then I stopped and went inside the house for a drink of water. I sat on the couch for a little while, then I got on my crutches and headed out again. Pat again asked me what I was doing, and I told her I was going to run my tractor again.

She informed me that she had to go pick up the twins after school in Hueytown, and I said "Fine, go ahead, but I'm not going back to Hueytown." She said she couldn't leave me by myself, and I told her I was not going back.

Then the Allison stubbornness in me really came out, and I told her to go ahead and go, that I knew how to make coffee and use the phone, and if I needed her, I'd call her. So she left me by myself that night.

About 7 o'clock the next morning the phone rang, and it was Pat. She said she hadn't heard from me and asked if I was all right. I told her she hadn't heard from me so what did that tell her.

Staying by myself that night definitely made a big difference in the whole healing process because I wasn't depending on anybody. A person with a head injury becomes very dependent on others, and as much as I loved my wife, I didn't want to be dependent on anybody.

I have to admit I broke the deal with my doctor once while I was rehabbing.

One day, I drove my four-wheel drive truck around in the pasture behind my horse barn, and I came to a wire gap I had there. I parked the truck, got out on my crutches and hobbled over to wire gap to close it.

Somehow I managed to get the wire gap pulled around to where I could almost close it, but it wouldn't go together all the way. The post had settled, and I wasn't strong enough to pull it together all the way to close it.

Well, I got mad, put both of my crutches together and tossed them like a spear at the truck. I turned around, walked over to the alley way of my barn, all the way to my tack room and got a piece of rope. Then I walked back and tied the gap up like I had to prove to myself it wasn't going to whip me.

At that point, I realized I didn't have my crutches, and I was a little bit worried about what I had done to myself. But I was so mad I didn't care. I just limped back over and picked them up, got in the truck and drove off.

I made my first post-wreck public appearance on July 21— less than a month after the wreck—when I spoke to a Kiwanis Club in Talladega. On August 2, I was still on crutches when I climbed up into the flag stand at Talladega and served as honorary starter for the 1981 Talladega 500.

I got back in a racecar to run competitively on Nov. 22, 1981, at Birmingham International Raceway.

I don't remember how I ran that day. I think I ran pretty good. But where I finished wasn't as important as just being out there racing again.

I do remember on my first lap that day, looking out the window and saying, "Thank you, God, thank you."

Thank You, Pat

When I got hurt in 1981, I can honestly say that if I had not had Pat Allison in my life, it probably would have been the end of me. She not only raised my children through a very tough time but also gave me the courage to get going again because man, I was struggling.

I'd sit in my chair at home, and somebody would stop by and say, "Oh, you look so good, Donnie." Well, I might have looked good, but I didn't feel worth a damn.

I was going through great depression, but I had no concept of what depression was. Through it all, Pat was right there by my side, making sure I didn't go off the deep end.

Looking back, the most amazing thing in the whole situation is that from the time of my first conscious moments in the hospital in Birmingham, my focus was on getting back in a racecar. There was never a question in my mind of not getting back in a racecar. The only question in my mind was *when* I would get back in a racecar.

DONNIE ALLISON

At the time, I never once thought about how much pressure I had to be putting on her because of the dangerous way I made my living. It wasn't selfish on my part, I don't think. It was just something I never thought about and Pat never brought up. Something we both accepted without words.

It had to be tough on her, but Pat has more backbone than any woman I know. The only one I would put in the same category with her is my mother, and that's some pretty high praise.

Whether it's been raising kids, racing, wrecking or recovering from injuries, Pat has stood by me through it all. (**Courtesy of Donnie and Pat Allison**)

I realize it took a pretty strong woman not to pack up and leave with what she went through over the years, though.

I was gone a lot, which left Pat to take care of the kids by herself. We never had any money. And there was always the chance she was going to get that phone call that said I'd been hurt bad at work ... or worse.

But to my recollection, she never once complained about being unhappy with our situation. She never once threatened to leave me. She never once asked me to quit racing.

That was probably the most important part of all because she knew from the time she first met me in Miami how consumed I was with racing. She knew I had made up my mind that was what I was going to do, and nobody was changing it, especially not some girl.

She was also smart enough after 20 years of marriage to know I was going to race again, no matter how badly I'd been battered at Charlotte, and she didn't try to stop me. She may have wanted to stop me, but she didn't, and that was important.

At the time, I was determined to get back in a car again, and that's all that mattered to me. I think I had to feel that way in order to accomplish what I had to accomplish.

Had she told me no, we would have been right back to the same situation I faced that day in my dad's shop when Bobby told me I'd never make a racecar driver. And it would have been me again saying, "I'll show you."

Pat knew what was going on with me. She knew me well enough to know that it would not have done any good for her to say no because, no matter what she might have said or done, I would not have listened to any of it.

If anything would have ever given us reason to separate, that would have been it, but I'm sure glad it didn't come to that because I love her better than anything in this world. There's never been anything that's come close, and there never will be.

Bobby Goes Airborne

I was standing on top of a hauler in the pits at the Alabama International Motor Speedway and didn't see Bobby have the wreck that changed NASCAR forever.

On May 3, 1987, I was spotting for Eddie Bierschwale, and at that time, they didn't have the nice, big stand they do now atop the track's press box. So we had to climb on top of our trailers and do the best we could from there.

At 2.66 miles, Talladega is so big, though, that you lost sight of the cars at different points on the track, and there was nothing you could do about it because you simply couldn't get high enough to see the whole track.

That's why I didn't see Bobby go airborne, flip over several times, almost take out flagman Harold Kinder, scare the life out of thousands of race fans and shake NASCAR to its core.

Bobby was going through the tri-oval headed for the start/ finish line when he ran over something and cut a tire. Somehow, the blowout shot Bobby's car into the air, then he spun around backwards and began bouncing.

Suddenly, it looked like the worst fears of everyone in the racing business were about to come true. A 3,600-pound car was out of control and heading straight for a grandstand packed with thousands of people. In a split second, it looked like a disaster of unthinkable horror was about to happen.

As I said, I didn't see any of this. All I saw over my shoulder was a big cloud of smoke rolling up from down in turn one, a long way from where I was. Ronald was on top of the hauler with me that day and he said, "Dad, that was Uncle Bobby."

I got on the radio to Bierschwale and hollered, "Turn one wreck, turn one wreck." Then I started looking for Bobby to come around, but he didn't.

Bierschwale got slowed down and got through the wrecked cars. As he did, he radioed back that it was Bobby and Phil Parsons and that it was pretty bad.

The next time through, he told me they were getting out of the cars and that the emergency crews were with Bobby. The next time by, he gave me some of the best news I had ever heard. He said Bobby was, miraculously, out of the car and standing up.

It truly was nothing short of a miracle that Bobby and a whole bunch of others weren't killed or hurt bad that day at Talladega.

When Bobby's front right tire blew, the force blasted the front of the car up into the air and the crosswinds caught him, spun him around and lifted him into the catch fence that surrounded the track. The catch fence was made out of chain-link fence reinforced with steel cables of various sizes, and it did the job that day.

Bobby took out over 300 feet of that fence and destroyed his racecar but only injured a handful of spectators. A few people were transported to hospitals in the Talladega and Birmingham area, and few others were treated in the infield care center, but no one was hurt seriously and, thank God, no one got killed.

NASCAR was forced to red flag the race for almost three hours to fix the fence, so we all had plenty of time to walk around the garage and talk to different people about the wreck. I was stunned, though, when I looked over at Bobby's hauler, which was parked right next to ours, not too long after the wreck happened, and there he sat on a cooler getting a drink of water.

He'd skinned the side of his face a little bit and the back of his left hand, but they had treated him in the infield care center and released him. I went down there, right up in the truck with Bobby, and said, "Are you all right?"

He was still pretty pale and shaken when he looked at me and said, "Donnie, you wouldn't believe what I just did. I took one heck of a ride."

That night, we all went home to Hueytown happy. Happy because Bobby hadn't been seriously injured or worse, and happy because Davey had gone on to get his first career Winston Cup win at our home track.

They taped the race, and I watched it and almost had a heart attack. When I saw what happened to Bobby, I literally broke out in a cold sweat. I looked at him and said, "Man, you did take a ride. You better believe you took a ride."

And all of those people sitting outside that fence in the grandstand took a scary ride, too.

What saved Bobby and those people that day was, plain and simple, divine intervention. There's just no other explanation for what kept that car from tearing through that fence and killing and injuring hundreds of people that day.

The ironic thing about the whole situation was that the track's insurance company had made them put extra cables in the fence all the way from turn four down through the tri-oval to the end of the grandstands in turn one just before that race. The folks that ran the track weren't happy about it, and they hollered about having to spend $300,000 or whatever it was to put the extra cables in, but it was a good thing because those extra cables are what kept Bobby out of that grandstand.

The cables were the massive steel ones they use on big ships to pull big barges across the ocean, and when Bobby hit, they held and bounced him back onto the track. Like I said, what else but divine intervention?

If Bobby had landed in the stands that day, it would have been a disaster for the fans, for him and for NASCAR—one from which it would have been very hard to recover.

Aside from the physical pain and suffering of those who might have been killed or maimed and their families, the psychological effect on Bobby would have been devastating. And NASCAR might still be trying to pay off all the lawsuits that would have followed in the wake.

Yes, a few inches of rounded steel cable and a few hundred thousand dollars saved everybody a lot of heartache that day and brought on perhaps the biggest technical adjustment in the history of stock-car racing.

The lasting effect of Bobby's big crash on NASCAR was that it ushered in the restrictor-plate era.

A lot of people probably don't remember it, but we had run with restrictor plates before, but they weren't being used at the time Bobby crashed at Talladega. As a result, speeds had climbed well over 200 miles per hour.

Bill Elliott had won the pole for that very race with a record 212.809 mph—a mark that will never be broken because NASCAR will never again let the cars run unrestricted at Talladega or Daytona. And the speeds in the 20- and 30-car freight train drafts they were running at Talladega were simply unbelievable.

It's very likely that Bobby, in the middle of one of those freight trains when he blew that tire, was traveling over 220 miles per hour, and that was simply too fast.

The fans loved the speeds, but they had crossed the line into the danger zone—for fans and drivers—so NASCAR had to do something. What they decided on was the restrictor plate.

Now those have become two of the most despised, most debated words in NASCAR racing.

Frankly, the drivers hate them, the engine builders hate them and the owners who have to spend millions of extra dollars each year testing and building cars just for four races at Talladega and Daytona certainly hate them. If the truth be told, NASCAR probably hates them, too, for all the controversy, griping and complaining they caused over the years, but they had to do something, and they feel like this is the best they've come up with so far.

I don't like them because I think there are other ways to slow the cars down, but I understand the reason for them.

I remember in 1978 Darrell Waltrip and I took Monte Carlos with the square noses, got together in a race and passed Cale doing 203 mph. That was 203 mph in a car which, aero-wise, was about like trying to drive a railroad boxcar around the track.

We got together in the draft, went to the front and ran five laps in a row at over 200 mph, and I'm telling you buddy, that's going pretty darn fast for a good stretch.

Somebody told me later that "Junior" Johnson got on the radio and told Cale that Waltrip and I were coming and for him to catch on to the tail of our draft as we went by. But we cruised by him so fast he never had a chance of hooking up with us in that draft.

I told Cale later that we went by him so fast it flip-flopped the car numbers on his doors, he just couldn't tell it because he was driving No. 11.

The principle behind the restrictor plate is to slow the cars down by cutting back on horsepower. You take the plate, which is a 1/8th inch-thick piece of stainless steel with four openings cut to whatever size NASCAR is allowing that particular year and put it between the carburetor and the intake manifold.

This limits the amount of air that gets into the motor through the manifold and "restricts" the amount of horsepower an engine can produce because the air and fuel don't mix in the right proportions within the engine.

For example, we probably ran 550 horsepower unrestricted when I was driving. Now they've got maybe 450 horsepower restricted. So even though the bodies are better aerodynamically, the engines won't accelerate as much, and that limits the top speeds they can reach.

Everybody hates having the restrictor plates because it makes everybody equal—drivers, engine builders and teams. Even after his big wreck, Bobby had the best saying about restrictor plates I ever heard.

He said in a press conference at Talladega one time that he didn't like the plates because it made other drivers as good as him when they weren't as good as him. He was right.

Restrictor-plate racing bunches up the field and gives drivers who may not have the skill or the experience and teams that may not have the best equipment a chance to run with the big boys on any given day, and that's not right. It's also dangerous because it makes drivers hold it wide open all the time, and that's why there's often a "big one" at Talladega and Daytona.

A driver running a restictor plate simply cannot back off the throttle for fear of losing the draft. In that respect, they take away a driver's control of his racecar and, for the most part, dictate his race strategy for him.

Without a restrictor plate, if you found yourself in a sticky spot in a race—say you didn't like the company around you—you could back off and get out of there. You could drop back to eighth, ninth or even further and when things sorted themselves out to your liking, you could catch up to the draft and go back to the front.

As a driver, I always wanted to go to the front if I had a fast car even if I saw something up there I didn't necessarily like. But if I didn't have a fast car, I'd just ride along, do the best I could do and try not to get aggravated at people until things up front looked better to me. And I wanted to make that decision for myself.

With a restictor plate, you can't do that. You're strictly at the mercy of the other guys. All you can do is ride along and hope somebody doesn't do something stupid which, unfortunately, happens all too often.

Instead of using the plates, I've always thought NASCAR should just limit the amount of lift allowed on the intake valve. In simple terms, that means they should limit how wide the air intake valve in the cylinder head can open and that would limit how much air would be allowed into the combustion chamber.

"Smokey" Yunick used to tell me that an engine is a pump, and it is. So if you limited the amount of lift at the valve and dictated the exact amount of air flowing into every engine, it would be a foolproof way to limit the horsepower—and the speeds—without limiting acceleration. You couldn't beat it, couldn't cheat it.

In fact, it would be as easy as stop and go for NASCAR to check.

It wouldn't be hard for NASCAR to make a big indicator with two colors on it—red and green—a needle for a pointer and a hose. Then as part of their tech inspection, the NASCAR inspectors could take the intake valve cover off, hook up the red-and-green gauge and rotate the engine one time through its cycle.

If the pointer went to red, that engine would be illegal. If it went to green, let'er go. It would work on the same principle of limiting the air flow to the engine as a restrictor plate, and all anybody would have to do to tell who was legal and who wasn't would be to look at the gauge.

So why don't they do it? The reason, to my way of thinking, is that it would take away NASCAR's control of the situation.

NASCAR, more than any other professional sports organization in this country, is in complete control of its own house. With the restrictor plates, they are the only ones who look at them and the only ones who measure them. They are in complete control, and that's not a bad thing.

Let's face it, NASCAR is the policing organization, so it has to have complete control. I'm a big advocate of policing. The only way you can control something is to police it, and NASCAR's gotten real good at it.

NASCAR has finally reached the point where they can make a rule and make everybody abide by it. If you don't like it, that's tough. It's the rule, and it's applied across the board.

There were times back in my day when I felt like some NASCAR inspectors played favorites in the garage. Now they don't. They don't have to, and I don't think they could get away with it if any of them even wanted to try it.

Remembering Davey

The hardest lap I've ever run in a racecar in my life came on July 25, 1993, at the Talladega Superspeedway. That day, I got into the No. 28 Texaco Havoline Ford to drive a pace lap around Talladega in tribute to my nephew, Davey, who had died 12 days earlier from injuries suffered in a helicopter crash at the track. It was one of the most overwhelming experiences of my life.

When they asked me to do it, I was proud because I was going to get to say goodbye to him in a special way, and he deserved it. But I had no idea how emotional it would be for me.

I cranked the car on pit road, and it hit me. All the good times and good memories of watching a special person grow from a child to a man. All the bitter sadness of his being taken from us too soon. As much heartache as if I'd lost one of my own.

As I drove around that big old track at Talladega, the people were all standing and hollering and waving, and I was waving, and it just overwhelmed me. The tears started and wouldn't stop, and before I was halfway around on that slow, slow lap, I almost couldn't see to drive because they flowed so freely.

That was a hard, hard time for all of us Allisons. We'd lost "Pop" in April of 1992. Bobby and Judy had lost their oldest son, Clifford, in a racing crash at Michigan International Raceway on Aug. 13, 1992. Then to get the news about Davey ... well, it was a rough 16-month stretch.

Clifford and I had been pretty close, because you always have a relationship with family, but nowhere near as close as Davey and I. Davey really was like one of my own.

He and Kenny were very close. They spent a lot of time playing together. When Davey was little, he played ball with all my boys, and later on, they hunted and fished together as much as they could.

I remember the boys used to tease him all the time because he'd get ready to go to the woods hunting, and he had this little fanny pouch that he carried crackers, candy bars, whatever he needed. They stayed on him all the time about it, and he just laughed and went right on like it never bothered him.

He was a good hunter and fisherman and as competitive at that as he was at racing. If he was in the woods, he wanted to shoot the biggest deer or the one with the most points on its rack. If he was fishing, he wanted to catch the biggest fish or the most fish. Whatever you did, Davey wanted to do it better.

He loved to come to the farm in Faunsdale. We had a bedroom there that was Davey's bedroom. He brought his stuff and left it there and fixed up "his" room at the farm just like he owned the place.

It was hard the day we had to go in and pack up Davey's stuff in that room, too.

Another thing I'll always remember about Davey was how hard a worker he was.

He came up the hard way. He was never given anything. He never had a silver spoon put in his mouth. He worked hard for everything he got.

I told the story earlier about giving Davey his first real race-car, which proved I saw something in him as a driver. So it will probably come as a real surprise that, at one point, I doubted Davey was going to make it in Winston Cup.

Early in his career, Davey's technical skills were a heck of a lot more advanced than his driving skills. Davey could do anything when it came to building a racecar. "Red" Farmer used to say you could give Davey two pieces of angle iron and he could build a racecar out of it.

"Red" was right, too, because Davey could build a motor and fine tune it. He could set up a chassis. He could do the work in the body shop, just about anything needed to build a darn good racecar ... then he'd go out and crash it.

Davey was trying to run the ARCA Series at that time in the early 1980s, and he was struggling. He was crashing all the time, and anything and everything that could go wrong did. To be honest, it made me start to wonder about him as a driver.

It worried Bobby, too. I distinctly remember a day at Darlington when Bobby came to me and said he wanted to go to dinner that night because he had something he needed to talk to me about. And whenever Bobby said he wanted to talk to me, it was about something pretty serious.

That night, it was about Davey. We were in the car heading to dinner when Bobby just came out and said he was concerned about Davey.

He was, too, because I remember we rode around Darlington until 1:30 or 2 o'clock in the morning talking about Davey and never went to get anything to eat.

Bobby asked me for my opinion and, of course, I shared it. I told Bobby that the best thing he could do with Davey was tell him to put his helmet under his arm and send him on down the road because he couldn't drive.

He said, "You really think that?" and I said, "Yeah, I really think that."

Man, was I ever wrong, and it was because I made an assessment without having all the knowledge I needed to make it. I didn't have all the facts, and I hadn't looked for them. I just said what I thought without checking all the details, and I was very definitely wrong.

I've told my brother that numerous times since then and admitted freely I'm glad I was.

How good could Davey have been had he not crashed that helicopter that July day in the Talladega infield? I think the sky would have been the limit.

I want to think he would have won a championship or maybe more than one. I want to think that he would have been right up there on the NASCAR all-time wins list with Petty, Earnhardt, Waltrip and his daddy. I want to think he would have been one of the best ever.

I think Davey had the makeup to be around the sport for a long time and be successful at it. I don't think he would have gotten burned out. I think with his personality and charisma, he would have relished the chance to play a role in NASCAR racing as it has evolved today.

I think Davey would have loved being a spokesman for a big-money sponsor, dueling Earnhardt, Dale Jr. and Jeff Gordon every Sunday and being in the spotlight. I know he would have loved winning a lot of races against the likes of those guys, and I think he would have, too.

The only speed bump for Davey might have been his health.

I know for a fact the bad wreck he had at Pocono in July of 1992—where Waltrip tagged him and he flipped 10 or 11 times—busted him up pretty bad. He had broken bones in his right forearm and wrist and a broken right collarbone, but he got back in the racecar the next week at Talladega and drove it to the first caution flag on raceday before letting another driver take over.

Davey had been flown straight from the hospital in Allentown, Penn., to the Talladega airport that Friday, had climbed in the car and found himself in so much pain he almost passed out. He finally decided to try it, though, after they figured a way to use Velcro to strap his right arm—which was still in a cast—to the stick shift and his left hand to the steering wheel.

It was an incredible show of toughness and determination on his part, but it may be have been something he regretted later.

He didn't give his arm the chance to heal like it needed to. Just like everybody else did at that time, he got back in the car too soon, and I know later it bothered him, not mentally but physically.

Davey would get out of the car after a race and have a lot of pain in his right arm and shoulder. Could that have affected him enough a few years down the road to force him to give up driving? There's no way to know for sure, but my guess is, just like that day at Talladega, Davey would have toughed it out for as long as his body would have allowed because he loved it so much.

The day I got the phone call saying Davey had died, I was getting some business affairs in order at home in North Carolina in preparation for a trip to Talladega the next day. I had just taken a new job with a driver named Roy Payne, and my first assignment was to go to Talladega with him to test.

Well, racing doesn't stop for births, marriages, deaths, nothing, so I went on down to Talladega, and for three days I ran that test looking at that helicopter hanging on the boom of a backhoe in the infield. Talk about hard work.

In the days after the crash, a lot of things were hard to understand and hard to live through. Now they are hard to reflect on.

I remember going to Bobby's house in Hueytown, going out back to his shop, putting our heads on each other's shoulder and crying plenty. I remember watching my mother kneeling in front of another casket and saying a whole rosary and never crying until she was through. I remember kneeling down in front of my nephew's closed casket and realizing I would never see him again.

It was one of the hardest things I've ever done in my life, and I almost couldn't take it. I almost can't take it now thinking about it again.

NASCAR: Then and Now

One of the worst things about my wreck in Charlotte was that it came at a time when NASCAR was beginning to really take off.

Some big money sponsors had started coming into NASCAR in the early 1980s, and it looked like plenty more might be on the horizon. Then I got hurt, and that pretty much ended any chance I might have had of ever getting with a top-notch team bankrolled by somebody with deep pockets.

After the crash in Charlotte, I was considered damaged goods, and I didn't help matters by staying away from the track like I did.

I got hurt in May and went to Talladega just over two months later to serve as the honorary starter for the Talladega 500. I even managed to get up into the starter's stand on my crutches. But for the most part, I stayed away, and that was a big mistake.

What I should have done was gotten back to the racetrack as quickly and as often as I could and let people see how I was progressing. At that point, I was 42 years old, busted up badly, recovering from a head injury and nowhere to be seen, which I'm sure led to speculation I was done as a driver.

I'm also sure the head injury was the thing that drove the final nail on my career because no big-time teams wanted to take a chance on an over-40 driver with brain damage, and I wasn't around to show them that everything—including my brain—was healing up pretty good.

As they say, out of sight, out of mind, and that was definitely the case for me after that wreck. It's a shame, too, because I'd like to think that some of the big bucks that came into the sport just a little bit later might have come my way.

I only had one "major" sponsor in my career and even that was misleading. "Hoss" and I hooked up with a guy named Ron Rice who owned a company called Hawaiian Tropic that made suntan products.

From 1977 until I left "Hoss" after driving three times in 1980, we ran 54 races with the Hawaiian Tropic colors on board. But our sponsorship deals with Ron weren't anything like the multi-million dollar ones they do today.

I say it was misleading because Hawaiian Tropic was a big company, and folks did come to associate us together on the track, but they didn't spend a whole lot of money with us to get that name recognition. In fact, Ron got a damn good deal.

We had begun negotiating with Ron in the weeks leading up to the 1977 Daytona 500, but as Speedweeks rolled around, we still didn't have a deal done. I went to one meeting with "Hoss" where we sat around with Ron and all his people, and he finally looked over at me and said, "Do you think I ought to do this?"

I looked at him and said, "I think you'd be crazy not to do this" because the deal on the table was $75,000 for the entire 1977 season.

Ron and his folks got up and went to meet in another room, and Hoss said, "What did you talk to him like that for?" I said, "Because he's crazy if he doesn't do it for that kind of money."

Well, Ron came back and said they wanted to go on the car for the Daytona 500 then talk again after the race about the rest of the season. So they agreed on that, and it was a good thing, too, because we were just about out of time.

Actually, they were so late getting the deal done that he only got the Hawaiian Tropic name on one side of the car because we simply didn't have time to paint it on both sides. Then I went out and sat on the pole, and he got his money's worth and much more in one day.

Ron couldn't have bought the advertising he got that day for $75,000, but he got it then and for 16 more races that season. If you put a calculator to it, it comes to $4,411 per race. Compare that to the $12 to $15 million-a-year sponsorships some of today's top teams command, and you can see how far NASCAR's come.

We didn't make huge salaries or run for huge purses, either.

It's ironic, but the biggest purse I collected for any NASCAR race I ever ran was the $39,600 I got for finishing fourth in the 1979 Daytona 500. The most I ever got for winning a race was $31,140 for winning the 1971 Talladega 500. And the most I ever earned in a single season was just over $142,000 in 1979.

Of course, all that got split with the car owner, so I have no way of knowing how much I personally pocketed out of those amounts. And in those days, that was good money—don't get me wrong.

But if you factor in inflation and put it in today's dollars, you'll see that the guys today are making those amounts many times over driving down the paths we paved for them.

People often ask me if I resent the fact that the modern drivers make so much money and have things so much easier than we did. My answer is no, not even a little bit because they earn it.

People don't stop to think that driving a Nextel Cup car is hard to do, and people who do hard things for a living should be paid well for it. You look at how close the competition is and how slim the margin is between winning the pole and going on home on any given day, and you realize these guys have to be on their game every day, every race. There's very little room for error.

The expectations placed on the current drivers are also much different from those that were placed on us.

These days, the sponsors drive the sport, and they have high—sometimes unrealistic—expectations. They also have plenty of pull with the car owners because they've got them by the pocketbook.

These days, sponsors want a driver to have a certain look and speak well because they want him or her to crisscross the country making corporate appearances. They may even want to build a television or magazine ad campaign around the driver. And, oh yeah, they want that driver to win races, too.

They're going to say, "We're paying you $12 million. We expect you to be in the top 10 all the time, the top five most of the time and to win as much as you can." And if drivers don't live up to all those expectations, car owners may have no choice but to make a change to please the sponsor.

It's understandable because everyone wants a good return on any investment, and they want it right now. But it puts a lot of pressure on a driver that wasn't there in my day.

There's also the issue of fan access and driver visibility.

In my day, we could go out to dinner with our family or to the grocery store just like any regular Joe, and it wouldn't cause a riot. Nowadays, guys like Dale Jr. and Jeff Gordon can't go anywhere without being mobbed by autograph seekers.

When we went to the track, we stayed in hotels and, occasionally, went down to the lounge or to a local watering hole for a drink just like everybody else. Nowadays, the drivers have million-dollar motor coaches they park in secured lots in the infield and spend most of their time in.

Two big reasons for that are convenience and economy, but it's also a fact that for many of them, it's the only place they can go to get away from the crowds while they try to do their jobs on race weekend.

Nowadays, a driver often can't walk from his car in the garage to his hauler or the hauler to his motor home without being swamped by fans who want autographs, pictures or just to shake hands.

It's a part of their job, and they have to do it, but what fans often don't stop to think is that it would be the same as Gordon or Dale Jr. walking into the middle of a meeting at their office and asking them to stop and pose for a picture.

Even more disturbing, and this one hits me personally, is that some "fans" are there just to get an autograph on a diecast or a picture or a magazine cover so they can turn around and sell it on the Internet. I know from firsthand experience that if a "fan" walks up and asks you to sign four or five items or more, those things almost certainly will end up on eBay.

They don't want my autograph just for a keepsake of our meeting. They want to make money using my name and have me do it for free. That bothers me, and I'm sure it bothers the current drivers.

I know for a fact it puts them in a tight spot from a public-relations standpoint because if they don't sign, fans think they are jerks, and if they do sign, they're often giving somebody who's going to turn around a make a profit off it something for nothing.

That's why many drivers today have adopted the sign-as-they-walk method. They don't stop for anything in the open garage area because they know if they do, a line will form, and nobody will be able to get any work done. So they walk and sign a few items and, to be honest, get the heck out of Dodge as fast as they can.

Nowadays, the drivers also are asked—or contractually obligated—to numerous corporate and charity appearances, many, many more than what we would have dreamed possible in my day. So a lot of the time we used to spend working on our cars and getting them ready to race, today's drivers spend flying around the country in their private jets, meeting and greeting.

It's a different kind of pressure, but pressure nonetheless, and it's what makes the racecar-driving life today so hectic. Sure, they make lots of money and have lots of toys, but the trade off for that is the loss of any semblance of a "normal" life.

They can't just jump in the car and go to the mall or McDonald's on the spur of the moment like we do. They often can't do anything without making arrangements for security first. Their schedules—and in great part their entire lives—are dictated for them by others. It's the price of fame.

I'll admit I wasn't the best at signing autographs and making nice with the fans, certainly not like Bobby. To his credit, Bobby would stand and sign and talk and pose as long as anybody wanted him to.

AS I RECALL...

I remember nights after races when we'd be all packed up and needed to get on the road because we had to be somewhere for a race the next night, but Bobby would still be standing there signing. It didn't matter if it was 1 o'clock in the morning and we had an eight-hour drive ahead of us. If folks were waiting on Bobby, he'd stop and sign.

I'd be like, "Bobby, we got to go," but he took care of the fans. If the truth be told, I should have done more of it, too.

I understand now how important that was—and is. It's so important that my advice to young drivers who have any designs on making it to Nextel Cup someday is to go take a Dale Carnegie course as soon as they can. Because these days, it's as important for a driver to be able to win friends and influence people off the track as it is to be able to drive on it.

In Good Hands

Talent-wise, NASCAR is in real good shape these days. There are more good drivers than ever before and more on the way. With guys like Gordon, Earnhardt Jr., Tony Stewart and the technology the way it is today, it's no stretch to say the competition to win a NASCAR race is as tough as it's ever been.

When it comes to the drivers out there today, Jeff Gordon is the best, plain and simple. He's the smartest. He's got the most ability. He's the All-American boy. I like the way he conducts himself. I just like everything about him.

He's everything you could want in a racecar driver, a P.R. person and everything else that goes with being a NASCAR superstar these days.

Dale Jr. and I are good friends, but I'd still look him in the eye and tell him he's a spoiled brat. He's a very, very good driver and very talented, but like a lot of those in today's younger generation, if they don't get their way, they go off and pout about it.

To my way of thinking, instead of Dale Jr. pouting about some of the things he does, he ought to stand up, take the lead and say, "We're going to do this or we're going to try that." But he's very reserved, and maybe it's understandable because of his position.

He followed in the footsteps of the biggest icon in racing history, and that's tough. Richard Petty got credit for being "The King," but if he was "The King," then Dale Earnhardt was "The Pope" of auto racing. And Dale Jr. will always have to live with that shadow of expectation hanging over him.

The sad part about it is that we might not see Dale Jr. ever realize his full potential as a driver because of that situation.

His daddy wanted to win races, and he didn't care whether he got paid or not for doing it. Well, Dale Jr.'s not doing it for the money, either, so he's got to want to succeed for other reasons, and that's where I wonder about him.

I'm not saying that to be derogatory in any way. I'm just being realistic in saying that Dale Jr. doesn't have to race to support himself or his family. He's already set for life financially and, for now, the main reason he races is because it's fun for him.

But if there ever comes a time when it stops being fun, when the demands on his time and the strain of being Dale Earnhardt's son start to weigh on him too heavily, it won't surprise me to see him walk away.

When you face all the circumstances that Dale Jr. has to face every day, there's a certain amount of intestinal fortitude you've got to have to whip them, and I'm not sure he's got it. I definitely think there may come a time when he gets tired of it all and hangs it up even though he may have plenty of good years as a driver left in him.

Another driver I put in the spoiled-brat category is Kurt Busch. He's got tremendous ability, but from the time he came into the sport, he acted a little bit too big for his britches.

The first few years he was in the sport, he didn't show much respect for the people who paved the way for him to get there, but it's definitely gotten better. That has to happen the longer you're around because you find out how hard it is.

It's not as easy as these young guys might think to come in and be competitive right away. Then they go nine or 10 months or a year even without even sniffing a top five, much less a win, and they start to have more respect for the guys who did it before them. That seems to be the case with Busch, at least.

Kevin Harvick is another driver who has as much ability as anybody and who finally seems to have gotten his deck of cards in order.

I think Kevin has had a little chip on his shoulder from the start, maybe because of circumstances. It was fortunate, but unfortunate, that he won as soon as he got into that ride with Richard Childress because it put him under the glaring spotlight right away, and he may not have been ready for it right then.

I was in Richard's truck that day before the Atlanta race in 2001, and I talked to Kevin before he went out and won that race. It was incredibly emotional for everyone involved, and I was proud of him that day.

I wasn't proud of him when I saw some of the things he did later, like the deal with the trucks in Martinsville in 2002 that got him suspended for one Winston Cup race. At that point, Kevin didn't realize that you don't go down to the courthouse and buck the sheriff or you go to jail, but I believe NASCAR has helped him have a much better understanding of it now.

Another guy out there whose name you hear a lot is Ryan Newman. He's a hard charger and, if the car is capable of sitting on the pole, he'll put it there because he doesn't show any fear.

Still, I'm not sure about Newman's racing ability. I'm not saying that he doesn't have the desire or the brains to race, but for some reason, I don't think he's got his book in complete order yet. I just don't think he always thinks about the right things at the right time.

Again, I'm not saying he's not a talented driver because he is. But I've said all along that somebody who sits on as many poles as he does ought to win a lot more races.

Of all the guys out there now, my man is Tony Stewart. Maybe it's because in a lot of ways he reminds me of myself, but Tony's definitely my favorite.

I talked to Tony quite a bit when he first made the move to stockcars. He even thanked me by name on TV in victory lane after his first win in Richmond, and that was special for me. What we talked about mostly was controlling—or not controlling—our emotions in certain situations.

Tony is made of old timey racecar driver stuff, and his attitude shows it. We had guys back when I raced, myself included, who if you asked them a question when they were having a bad day, would tell you in no uncertain terms to go jump in a creek. Tony would do that, too, which may have been what reminded me so much of myself.

I know for a fact that he felt like he was done wrong by the media early on in his NASCAR career, and that's something we could talk about because I felt the same way at times in mine.

I took personal some of the things the media said about me. What other way are you supposed to take them? I knew it probably wasn't the right thing to do, but it aggravated me, especially when I didn't have a newspaper or a magazine to put my side of it down in like they did.

The only way I had to retaliate was when they stuck a mike in my face or asked me a question, I'd say something smart. When

I did that, though, I just got my ass in the crack further. Well, Tony's been in a couple of pretty big cracks himself for just the same reason.

There was the situation he had with Darrell Waltrip a few years back where Waltrip criticized him during a Fox broadcast, and Tony later responded with a backhand shot at Darrell. Soon the two were trading insults on the air, and there was no way for Tony to win.

That's what I told him, that he had to concede he couldn't win in that situation. Darrell had four hours to talk about Tony, and whether he was right or wrong in his assessment, Tony wasn't going to change anybody's mind in the few seconds of airtime he got before or after a race.

It was frustrating, I know. Somebody says something about you and you want to answer back, but you don't have a real opportunity. It's only human nature to want to defend yourself, but you can't, and that makes it even more frustrating.

Then you put it in the context that racecar drivers operate under, and it only frustrates you even more. You've been in a blazing hot car for hours, you're frustrated by something that's happened in the race or maybe you're mad about something that didn't happen like it was supposed to.

You climb out of the car and immediately someone walks up, sticks a microphone in your face and says, "So-and-so said so-and-so about you. What do you think about that?"

You don't know exactly what was said. You don't have all the facts. But you're tired, hot and angry, so you get hacked off and say things you shouldn't in response.

You know you shouldn't do it. You know that if you could have a face-to-face conversation with that person under different circumstances you might not say anything so combative—you might not even say anything at all. But in the heat of the moment, you unload on somebody and come off looking bad and making enemies.

That's what Tony did early in his career, and he'd agreed with me on that. But he's gotten much, much better about it over the years, and that's made him even more of a complete driver.

In terms of ability, he's as good as Gordon. In fact, if I ever needed a driver to drive my car in a race I absolutely had to win, Tony Stewart is the guy I'd want.

Past, Present, Future

The kind of call all successful athletes hope to receive as the capper to their careers has actually come to me twice over the last several years. The call that says you are officially recognized as one of the best in your sport. The call from the Hall of Fame.

The first in late 2008 was to tell me I'd be inducted into the International Motorsports Hall of Fame in Talladega with the Class of 2009. The second came in late 2010 with the news that I had also been elected a member of the Motorsports Hall of Fame of America in Detroit and would be inducted in the Class of 2011.

Both sounded as sweet to my ear as a finely tuned engine.

The call from the IMSHOF was particularly satisfying for a couple reasons. First, I always considered the Talladega Superspeedway my home track, and the IMSHOF building is right there on the speedway grounds. Second, the NASCAR Hall of Fame was not yet up and running, so at that time getting a call from Talladega was the highest honor anyone could receive in motorsports, and the call from Detroit ranked right up there with it.

The thing that made my selection to these prestigious institutions so gratifying was their diversity. They are very similar in the diversity of people inducted into them. They are made up of the best racers from every corner of our sport and every era. So for those institutions to acknowledge what I did in my career, for them to say "you are one of us" is an honor extremely difficult to put into words.

I am truly blessed in the number of organizations that have chosen to recognize me for the impact I had on my sport and the sports world. I was inducted into the Alabama Sports Hall of Fame in 1999, the Florida Sports Hall of Fame in 2000, and numerous other racing halls along the way, and I am grateful for each and every induction.

It amazes me when I think about all the honors I received over my career, and even more so considering that much of what I accomplished I did so while living with a serious disease I never knew I had until years later.

When I came back from Daytona to Alabama in 1970, for some reason, I could not get enough to drink. I have never been one to stop much when I am on the road, but I distinctly remember that on that trip Pat and I stopped three or four times just so I could get water. I didn't think much about it because it seemed like every year around Daytona time I got a bad cold, and I just chalked it up to that. But for some reason I just couldn't shake it.

Well, "Banjo" and I didn't go to the race at Richmond the next week, but when I rolled into Atlanta a few weeks down the road, I was still sick as a dog. That was OK, though, because I knew all the folks in the infield care center at Atlanta and knew I could get a shot there.

It seemed like I always had to get a shot of penicillin or something every year to finally make me feel better, so I kept telling myself, "Man, I can't wait to get to Atlanta and get me a shot so I can get over this."

At that time, we started practice on Thursday morning, so I ran the car and then went to the infield care center to see the nurse, Linda, who I had known forever. I told her I needed a shot because I was sick, and she told me to lie down and she would get the doctor.

When he came back, there was an Oriental lady doctor with him, and I distinctly remember that she walked up, took one look at me and said, "Look how jaundiced he is." Well, I didn't know what the hell jaundiced was, but I do know that when the track doctor pressed on my stomach, I almost jumped out of that bed.

They quickly sent me to Grady Memorial Hospital for tests, and when the doctor who ran them came back, he immediately asked, "Do you have family with you?" I told him my wife was coming from Alabama that night but that right then I was alone. Well, he wanted to check me into the hospital immediately, but I protested. I said I had to go qualify a racecar, and he said, "Son, you can't qualify a racecar. Your liver is about to burst."

That shook me up enough that I promised him if he wouldn't make me check into the hospital I would go back to the motel room and when Pat got there I'd go home to Alabama and see my doctor. They got Pete Hamilton to drive the racecar, and that's what I did.

My doctor was pretty shook up, and he wouldn't let me do anything. He told me to go home, get in bed, drink milkshakes, eat steak and suck on hard candy, and buddy, I ate enough hard peppermint to fill a dump truck.

His prescription worked, and I started feeling better pretty quickly, so I went back to racing and never really thought anymore about it. After that initial episode, I never hurt. I never had another flare up like that again. Nothing.

It honestly never crossed my mind again until August of 2000 when Pat decided we needed to get a different life insurance policy because a policy for a racecar driver was, for obvious reasons, very expensive.

She had been after me for awhile to get another policy, but I drug my feet. I told her that I didn't want to do that until I retired from racing and, at that point, I still did not consider myself retired even though I wasn't running anymore. But it finally reached the point where I told her to find out what we had to do to get a new policy, and I'd do it.

So the insurance company sent a lady to our house to do a physical, draw blood, the whole nine yards. In about two weeks, they sent a letter to my primary care doctor saying that I had been denied the policy because the tests showed I had Hepatitis C. Of course, the doctor did his own tests and, unfortunately, the next call I got from him was to confirm that, yes, I had Hepatitis C.

I was stunned. I wasn't sick. I didn't feel bad. I had given up drinking alcohol in 1998 because I wanted to, not because of my liver or anything like that. So I was suddenly in a fight for my life, but I didn't come out swinging right away because frankly, even with the warning, I still didn't take it all that seriously.

In fact, it was early 2004 before I finally went to a liver specialist in Charlotte, who ran his own set of tests that once again confirmed the positive HepC diagnosis.

When I first met him he had been joking around and cutting up with me. I remember he stuck a needle in my side for a test and laughed and told the nurse, "Look, he doesn't even bleed." When I went back to his office, he wasn't joking anymore. He said the tests showed that on a scale of 1 to 4, with 4 being Cirrhosis of the liver, mine was at 3.

He looked me in the eye and told me that left undetected I would have had Cirrhosis of the liver in five years and might not have ever known it was coming on until it was there. He said, "There are two cures for Cirrhosis of the liver. One is a transplant and the other is a pine box." That got my attention right there.

Fortunately, I was healthy overall, which made me a good candidate for a clinical study being conducted by two pharmaceutical companies to determine who had the best HepC medication.

The liver specialist gave me a notebook to read about the pros and cons of the study. One pro was that if I participated, my medication would be free. If he prescribed it, it would cost $30,000. But what it came down to was I really had no alternative but to participate.

If I did, there was a chance I could lick it. If I didn't, the HepC was going to win anyway. So I signed up for what I decided was the lesser of two evils.

What followed was a routine that left me never wanting to see a hypodermic needle again if I can help it.

For the first 12 weeks, I went to Charlotte every Monday to get a shot and give blood. Then for the next 48 weeks, I gave myself a shot every Monday and went to Charlotte every month to give blood. That was tough, but even tougher was the fact that throughout the whole program, for over a year's time, never once did they volunteer any information about how I was doing.

One time I asked the nurse, a wonderful lady named Martha, how things were going, and all she said was, "Your blood count is looking better."

It was tough because in my mind, my clock was ticking. This program was going to take one year of my five-year allotment and, if it didn't work, what then?

But on March 19, 2006, I got the kind of news that makes you drop to your knees and say, "Thank you, Lord!" It was ironic because that was my dad's birthday, but I was the one who got the present that day.

Most times when I went, the doctor would come in a minute or two later just to acknowledge I was there. But that day, he didn't come in for 5 minutes, 8 minutes, 10 minutes, and Pat and I were sitting there thinking the worst.

I asked Martha where he was, and she said he was in another room with a lady who wasn't as lucky as I was. About 10 minutes after that, he came bouncing around the corner hollering, "You're cured! You're cured! You no longer have Hepatitis C!"

Upon considerable reflection, I believe I contracted the disease when I wrecked a super modified really bad in Laurel, Miss., in late summer 1969, collapsed a lung, and had to have a blood transfusion. But it doesn't take much thought at all to know I am truly lucky to have beaten it, and I have a letter from my doctor, a note from Martha with a checkered flag at the top that reads "Well done—race won" and a life insurance policy to prove it.

The one racing hall of fame I am not yet a member of is the NASCAR Hall of Fame in Charlotte, 30 miles south of my home in Salisbury, and I admit I very much want to be in that one.

Bobby was one of the many great racers who attended the ceremony in Detroit to welcome me into the Motorsports Hall of Fame of America when I was inducted in 2011. (**Courtesy of Mark Scheuern/MSHFA**)

I have been fortunate to attend every induction ceremony so far. I have participated in several other events there and done any promotional work the staff has asked me to do. It is a magnificent place and, with all due respect to other racing organizations or football or baseball, I don't think there's but one better, and that's the one the good Lord runs.

NASCAR's promised land is a 150,000-square-foot entertainment center that includes more than 40,000 square feet of exhibits and artifacts. Among those are several that chronicle the racing Allisons in general and my career in particular.

None strikes me more than one located on Level 4, right across from a life-sized statue of Bill France Jr. There you'll find full-size replicas of the cars in which Cale and I crashed our way to racing infamy in 1979 along with a full-length video detailing the race and its historical significance.

Every time I go there, I'm struck time and again by the sheer magnitude of that event and by an incredible sense of irony. It's ironic because they chose to put this big statue of Bill Jr. right across from our display as if he's sitting there in a chair constantly watching and saying, "I know what you guys did to make all this possible."

One time before he died, the subject of me not being in the IMSHOF came up, and he looked me straight in the eye and said, "I don't know why you are not there yet." It was nice of him to say, and I felt the same way, but I hope no one ever has to say that to me about the NASCAR Hall of Fame.

I'm never going to actively campaign for it because I just don't believe in that. But I do believe that if you add up what I accomplished on the track and what I contributed to the sport off it, the total should be enough to earn me one more of those sweet-sounding phone calls someday.

If that call never comes, I won't be devastated because I know there are a lot of people who did as much or more for NASCAR who may never get in, either. What would devastate me, though, would be to see the Allison racing legacy end without me doing all I can to prolong it. To me, the Allison name is no different than Petty or Earnhardt ... it is synonymous with NASCAR racing, and I want to see that continue.

For several years now, I have tutored my grandsons Taylor (Pam and Hut's son) and Justin (Kenny's son), in the art of auto racing. Taylor still dabbles in it some, but as he has gotten older, his focus has turned more towards running the family business.

The one who has the racing bug now is Justin, and I truly believe he has the potential to take the Allisons back to NASCAR greatness someday.

I guided Davey when he was just getting started. More recently, I mentored young drivers like Joey Logano and Trevor Bayne when they drove our Allison Legacy Series cars. I saw something in all of them that told me they could make it, and I see the same things and even more in Justin.

He's got a lot of his Paw Paw in him. He's got a little bit of a temper just like me. He's got an uncanny ability to miss wrecks

I was very proud to have my grandsons Taylor Stricklin, the tall one on the left, and Justin Allison with me at my International Motorsports Hall of Fame induction ceremony. It represented the old and the new of the Allison racing family. **(Courtesy of Jimmy Creed)**

just like me. I've seen times when I just knew he was going to wreck, but he didn't, and how he missed I have no idea. And man, can the kid drive … just like me. He's progressed from go karts to our Legacy Series cars to late models, and he's been competitive at every level.

He won a Legacy Series championship a few years back because he never finished lower than third in 20 races. He got in a late model for the first time ever at Hickory Speedway, a very fast .363-mile racetrack, and on his third lap drove the car as fast as the owner ever had. He's definitely got what it takes to make it big if he works hard, listens and learns. So I've told him I will do all that I can to help him get there and then it will be up to him to run with it.

I have all the confidence in the world he can do it, but for us to see an Allison driving at Talladega, Daytona or Charlotte again, several things have to fall squarely into place. He has to stay focused on racing. He has to continue to develop as a driver. And he has to find the right team to go with from the start.

Racing is big business, and there are those out there who wouldn't hesitate to capitalize on the Allison name to improve their particular situation. We've already been approached to put Justin in some ARCA cars and even some Craftsmen Truck Series rides, but the situations were just not right.

I just felt those folks were looking to grab whatever bang for the buck they could get from having the name Allison on the side of their ride, and they weren't concerned about Justin's overall development.

We want to see Justin in an ARCA car or a truck or a Sprint Cup car someday more than anybody else out there, but we want it to be in a spot where he has a chance to run hard and well for the checkers from the start.

If he gets out there in a bad ride and things go wrong, folks wouldn't blame the owner or the crew chief. They would just say, "Justin Allison can't drive." But, trust me, that's not the case.

So even though he's a little later arriving on the scene than some of the young hotshots we've seen over the last decade or so, we're trying to be patient, take it slow, and put him in the right ride from his first green flag.

The old racecar driver in me wants him to heed some of my lessons from the past and avoid making the mistakes I made. The grandfather in me wants to see my grandson get all the breaks he possibly can and be the best he can possibly be. The Allison in me desperately wants to see the day when once again we gather as a family in Victory Lane to celebrate a win on NASCAR's highest level.

When that happens—as I truly believe it will—the circle of my career will have closed completely in the most fitting way any racer could ever dream of.

Final Thoughts

When NASCAR released the list of its 50 greatest drivers as part of its 50th anniversary celebration in 1998, my name wasn't on it. Bobby's on the list. So are "Red" and Davey. But I wasn't on it, and even Bill France Jr. himself said that was B.S.

It hurt my feelings bad when the panel of "experts" that voted on that thing didn't include me, especially when there were drivers on the list I know I was better than. That's not saying they weren't good drivers, but I beat them a lot more times than they beat me, and they got in. So why not me?

I believe it's because people don't realize how good a driver I was, and a lot of the blame for that falls on me. I didn't believe in talking to the press back then. I viewed it as tooting my own horn, which was something I hated seeing others do.

I felt if I did my job and won the race that spoke for itself. If I didn't do my job and didn't win the race, that spoke for itself, too. Either way, I didn't feel like I had to explain anything to anybody.

What I never stopped to think about at the time was that I could have helped people appreciate my abilities more—and cleared up some misconceptions about me along the way—if I had been more open.

People have often asked me why I never really got a shot to drive for one of the top-notch factory teams. One reason is because I never made the effort to let any of them know that's what I truly wanted to do.

There are times in a man's life when he has to put forth an effort to get what he wants. You've got to go to somebody and say, "I want to do that." I didn't do that and that was a big mistake on my part.

None of the factory teams ever really solicited me, but then I never solicited them, either. If I had, there's no telling what might have happened. I might have flopped, but I'll bet everything I own I wouldn't have.

If there's anything I regret about my racing career, it's that I didn't take the initiative to solicit me a good, high-caliber team to run for a Grand National or Winston Cup championship. If I had, there's no doubt in my mind I would have won it. But even if I hadn't won a championship, everybody would have damn sure known I was there running for it.

I won championships at every racetrack I ever raced on in my short-track career. Dixie Speedway, Montgomery Speedway, BIR, Chattanooga, all of them. I won hundreds of heat races and features. I was rookie of the year in the Grand National series and Indy. I guess I just felt that maybe somebody should have come asked me to drive for one of the factory teams because I did all that winning, but they never did.

If there's anything else I regret, it's that I didn't take more time to openly smell the roses while I was racing, especially after I made it to the Grand National/Winston Cup level.

This is the car I drove to Victory Lane in the World 600 at Charlotte in 1970. I will always be grateful to "Banjo" Matthews for giving me the chance to drive it. **(Courtesy of Donnie and Pat Allison)**

I ran 242 races in my career, not a lot compared to Petty's 1,184 or Bobby's 717, but 242 more than a lot of other drivers who would have loved the shot but never got it. I won 10 races and 17 poles. I started top 10 in 163 of those 242 races and finished top 10 in 114. That was 63 percent of the time starting in the top 10 and 47 percent of the time finishing there.

I got to race on some of the world's greatest racetracks. Daytona Beach. Talladega. Charlotte. Atlanta. Bristol. Martinsville. Darlington, which was my favorite. I got to meet dignitaries and celebrities and travel the country. I made a good living, met a lot of good people and had a good time. It makes me a little sad now that I didn't enjoy it more while it was happening.

At the time, though, I was just intent on winning, and that made me narrow-minded. I couldn't see that it was OK for me to tell people I was good in a racecar. That sometimes you have to explain things so that folks truly understand and appreciate your talents. That you can say, "Yeah and I did it, and I'm proud of it" without it being bragging.

I'm proud of what I accomplished in my career. I'm proud of being someone who gave 110 percent. Of being a driver who, no matter what kind of car I was in, always ran up front and had a chance to win. That I was someone who didn't get out there just to make laps and collect a paycheck.

Finally, I'm proud I got this chance to share these moments and memories as I recall them. I hope they'll help those who knew me back then understand me as person and those reading for the

"Banjo" Matthews wasn't just one of my car owners. He was a teacher and a friend and one of the most knowledgeable racing men I ever met. (**Courtesy of Donnie** and Pat Allison)

first time appreciate me as a racecar driver who got up on the wheel with the best of them. I know it's helped me appreciate myself.

It's like "Banjo" told me very late in his life. We were at his house, and he was sitting there in a wheelchair with an oxygen tube in his nose, dying. As the conversation came to a close, "Banjo" took my hand, looked me straight in the eye and said, "Donnie, you were a helluva lot better racecar driver than you ever gave yourself credit for."

As usual, the old "Banjo" man was right.